TWO GULLS
AND A GIRL

A Study of a Seagull Colony
by a 10 Year Old Naturalist

ROXANNE SCHINAS

Illustrations by **Roxanne Schinas**
Photographs by **Jill Dickin Schinas**

IMPERATOR
PUBLISHING

Published by *Imperator Publishing*
109 Esmond Road, Chiswick, London W4 1JE
www.imperator-publishing.com

Text and illustrations copyright © Roxanne Schinas 2008
www.yachtmollymawk.com

Photographs copyright © Jill Dickin Schinas 2008
www.jilldickinschinas.com

First edition published in Great Britain in 2008

ISBN 978-0-9560722-0-7

A catalogue record for this book is available from the British Library.

Printed and bound in Great Britain by *Lightning Source UK Ltd*
Chapter House, Pitfield, Kiln Farm, Milton Keynes MK11 3LW

Contents

Foreword *by Richard Williamson* 5

Introduction *How It All Began* 7

Chapter **I** *Kidnapped!* 13

Chapter **II** *Baby Birds Aboard* 25

Chapter **III** *Remus Lives to Tell the Tale* 33

Chapter **IV** *Swimmers and Bombers* 47

Chapter **V** *Flying Lessons* 59

Chapter **VI** *Gull Talk* 71

Chapter **VII** *Feeding Habits* 85

Chapter **VIII** *Murder Mystery* 99

Chapter **IX** *Leaving Home* 109

Chapter **X** *Island of the Insects* 117

Epilogue *The Return of Romulina* 133

Foreword

Roxanne has some rare gifts. The first is that she is fascinated by the natural environment. Most people have only a passing interest. Without this, most of planet Earth's riches are locked secrets from the human race. You are born with this gift and it provides a lifetime of enjoyment unknown to others.

Her second, more precious gift is that she is a worker, with energy. So she will be useful to the planet too. She may become a scientist, in which case her diligence and integrity should result in valuable achievements.

The first gift feeds the second of course. Even so, there are millions of sad cases throughout history where these gifts whither without the good fortune to have an encouraging environment. Parents, partner, even the luck to be in the right place at the right time. So far everything fits into place for Roxanne. The result is this book. It is compelling because her enthusiasm is compelling. She can tell a story too, and it is a truly delightful story, quite wonderful. My father had these gifts and most of these advantages. The result was *Tarka the Otter*, among sixty other books.

The sky is the limit for Roxanne. Bon voyage.

Richard Williamson
October 2008

Introduction
How It All Began

The ripples from the dinghy's bow shot ahead of her across the calm water as she was pushed forward by Daddy's powerful stroke. I crouched in the bow next to Poppy and called out to the others. Xoë waved back.

As we got closer Poppy braced herself for the jump, and then just before we arrived she leapt from the boat onto the shore, soaking her hind legs with salty water. We all laughed.

The dinghy touched the beach with a gentle scrunch and I jumped out onto the sand, as eagerly as the dog, splashing my ankles with the cool water and jumping onto the beach. After Mummy and Daddy had clambered out, Caesar and Xoë pulled the boat up onto the beach next to their Topper.

It was Mummy's birthday and we had spent the morning anchored at the northern end of the Mar Menor, watching a spectacular airshow. Then my brother and sister had set off in their little sailing dinghy, bound for Isla Perdiguera. Caesar was fifteen; Xoë was thirteen. I was only eight, so I was not allowed to sail far away with them in the Topper. I had to stay with Mummy and Daddy on our yacht, *Mollymawk*. The Topper had set off first, so it had reached the island first, ahead of *Mollymawk*.

"There are chicks at the top of the big hill," Xoë babbled, in a hurry to be the first one to tell of the discovery.
I looked up at the hill. Like the rest of the island it was barren and dusty, covered in dried branches and bushes, dead and brown.
"They're ever so fluffy," my sister went on, "and I think they're gulls. Let's bring one home as a pet. Do let's! You'll let us, won't you, Mummy?"
"No."

We ran up the hill, tripping on the rocks and stones and sending them bouncing down into the bushes below. My parents walked after us, Mummy all the while taking photos and telling us not to scare the birds off their nests - and saying, over and over in response to Xoë's nagging, "No. You definitely can't have a chick."

Poppy pulled at the lead, excited by our running and by the screams of the seagulls above us. As we got higher up the hill the gulls mobbed us, trying to scare us away from their nests and their young. Then we found the chicks. Not really chicks at all, but immature birds near to fledging.

Mummy took photos of the gulls, the chicks, the flowers, the sea, and everything else in sight. She kept wanting us to pose near a nest, or hold a branch out of the way, and all the while she kept saying to Xoë that she could not have a chick. And Xoë kept on pleading, "Just a little one. Only a baby."

At last we got to the top of the hill. We could see all over the whole island and the edges of the Mar Menor were visible in a big circle around us. The breeze refreshed us after our hot climb, and we plunged down the other side of the hill, slipping on the clattering rocks and scratching ourselves on the bushes. Then we went up the next hill. In the side of this one we found a cool dark cave. There was a pin prick of light at the other end.
"It's a tunnel! It goes right through the island!"
We passed through the tunnel. And, at last, having explored everywhere, we went back, past the hills and the bushes, past some old war time ruins, to our dinghy hauled up on the beach.

That first visit to Isla Perdiguera was more than three years ago. Since then we have visited it almost every birthday and at Christmas, and whenever we felt like it. We have come here in winter and found everything surprisingly green and well-watered, and we have come here many times in the summer and found everything dead again.

Every time we come to Perdiguera we find something new. A seagull's feathers; a snake skin; a cicada grub's carapace. A huge pen shell, washed up on the beach; an old warbler's nest; a kestrel's skull; a mantis's ootheca. We always find something interesting. It is a wonderful place for me, as I love wildlife and we always find something new to add to my collection.

Sometimes we find something living. Once it was a rabbit which we found hanging from its back leg in a thorn bush. We rescued it, but its leg was injured, so we hid it in a bush with plenty of cabbage leaves. By the next day the leaves were gone and so was the rabbit. It must have got better.

Another time we found a seagull which was sick, and we brought it back to the yacht and fed it. Another time we were walking along the beach when dinner washed up at our feet, in the form of a huge, live mullet! It had been thrown ashore by a motor boat's wash.

But all the time what I was really hoping to find was a poor little abandoned chick...

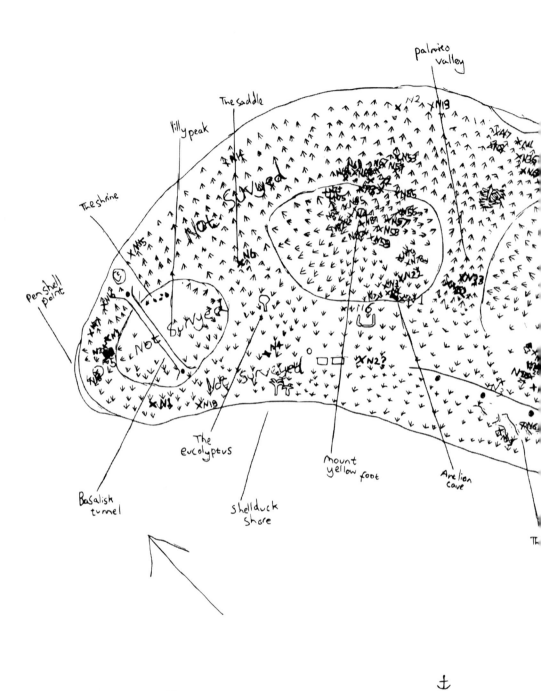

Isla perdiguera

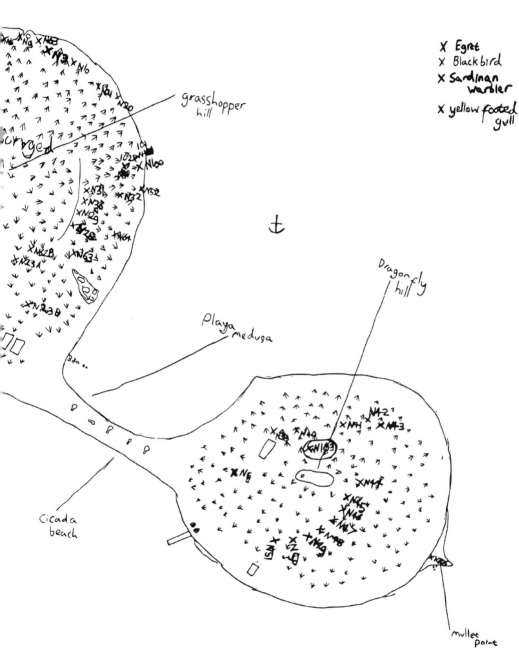

I
Kidnapped!

On the Mediterranean coast of Spain just near to Cartagena there is a "little sea" or Mar Menor. It is about fourteen miles long and six miles wide, and it lies trapped behind a long spit of sand which is covered in tower blocks. The salty lake is warm and calm, so it is very popular with tourists, and the tower blocks are all filled with holiday apartments.

There are two islands in the Mar Menor. The biggest one is called Isla Baron and it is privately owned – by a baron. It is said that wild goats live on this island, but I have never seen any. Every weekend in the summer visitors come from the tower blocks in their speed boats and anchor off Baron, but they cannot go ashore. Nobody is allowed to land here.

The other island is my island – Isla Perdiguera. The tourists come here, too, and they come ashore, but most of them never go further than the beach. The average visitor probably thinks that the island is completely dry and empty – but actually it is covered in life. Anyone who looks closely will see the rabbits running into the grasses, and hear the little Sardinian warblers scolding from the bushes. Any who are curious will turn over the stones and find millipedes curled up tightly in the dusty earth. You must look very closely to see the ants and the spiders crawling in the bushes, looking like the leaves and twigs that surround them. But there is one animal that even the most ignorant and unobservant visitor could not fail to see, and that is the gulls. The gulls soar over you in the pale blue sky, screaming and chattering amongst themselves, or they sit on the rocky outcrops and the ruined buildings, looking smart and business like. They rule the island.

There are several types of gulls in the Mar Menor. There are *Larus auduinii*, the rare, endangered Audoin's gulls which nest on an island just outside the entrance. There are a few black-headed gulls, and there are some slender-billed gulls. But the commonest gulls by far are *Larus cachinnans*, the yellow-legged gulls. (Some books call them *L. michahellis*.) The yellow-legged gulls have a white belly, chest, and head, grey wings, black and white tail feathers, a yellow beak with a red dot, and - of course - yellow feet and legs. They look very similar to herring gulls, and they are so closely related that until a few years ago they were considered to be the same species. In fact, the only obvious difference is that the adult yellow-legged gulls have yellow feet and legs, whereas the adult herring gulls have pink.

The yellow-legged gulls are the ones which nest on Isla Perdiguera. The island must be heaven, as far as they are concerned, for they could not find a safer place to breed. Apart from humans there is almost nothing here to hurt the eggs or the chicks, and even humans rarely go walking round the island. So, as a result, the population of gulls has grown huge. The island has hundreds of nests on it.

Since the island cannot possibly support such a large population the birds must have help - from humans. The humans are not actually interested in helping the gulls, but the gulls have learnt to take advantage of our activities. They fly along after fishing boats, and they visit the rubbish dumps on the mainland. Although they would probably prefer fish, and it would definitely be better for them, it is easier for them to find food at the dump. Once we took apart something that a seagull had sicked up, and we found it was a piece of raw beef steak!

There are always gulls hanging around at the island, and in the spring hundreds of others arrive and they all nest there. Their nests are not pretty woven things like a song bird's nest. They are just little mounds of grass and leaves, with a dimple in the middle. First the birds scrape a hollow in the ground. Then they line it, usually with grass and leaves which are piled up above the level of the ground and shaped into a shallow bowl. They must shape the nest by sitting in it, I suppose. Then they pluck out a few feathers to make a soft bed. The nests which have the most feathers are always near to a dead gull!

An ideal nesting site is sheltered by a bush. Sometimes it is almost under the bush. The bush hides the nest and it also gives somewhere for the chicks to shelter from the sun. I think that the best place for a nest is on the shore of the island, and this is where many of them are. It seems that the closer the water the more valuable the nest site is. The nests by the shore are almost always sheltered by bushes, and sometimes the bush is almost overhanging the water. The nests are always on the side facing away from the water.

A typical nest, beside a bush.

I found the first nest on March 9th, but the birds had probably been making them for at least a week before this. I found the first egg on the same day. This was very early, and I know that some of the gulls were still laying eggs at the beginning of May.

The eggs are greenish brown and they are about the size of a chicken's egg, but much more pointed at one end. Some are darker than others, some are browner, and some are bluish. They are covered in blotches, speckles, and spots of a very dark brown colour, with some pale greyish blue spots. The colour and pattern varies a great deal. It probably depends on what the mother was eating.

When I came back again on March 10th there were two eggs in this first nest, and on the next day there were three lying point to point in the nest. Soon I had found several nests, and I decided to make a map and mark all the nests that I found. I decided to make a study of the nesting colony on Isla Perdiguera.

The nest usually contains three eggs.

The mother yellow-legged gull nearly always lays three eggs, but we had been told that only one chick from each clutch would survive. This was one of the things we wanted to observe, to see if it is true. When I began my study, in April, there were three tiny fluffy chicks hatching in each nest. But when we were here in June of the previous year there seemed to be only one large feathered chick at each nest. The other two had apparently died.

We had been told, by various people, that the parents cannot find enough food for three huge chicks as big as themselves, and so the strongest chick gets all of the food that they bring. Presumably it would be the first chick to hatch who would be strongest, as he would have a head start. His siblings would starve. So my sister and I decided to save one chick. On the day that we found that first nest and the first egg we made a plan...

The idea of adopting a seagull chick was actually given to us by one of the nature wardens from Isla Grosa, the Audouin's seagull colony. The nature wardens don't like the yellow-legged gulls, because they say that they compete with the Audouin's. The previous year, in spring, one of the wardens gave me a yellow-legged gull's egg.

"You can hatch it and rear it," he said, "or you can have it for your breakfast."

We put the egg under a lamp, and for a month we kept it at the right temperature, but the egg turned out to be addled. Perhaps it was already cold before it was given to us, or perhaps we didn't turn it often enough. This time we decided to take the easy route. This time we would let the gulls do the work of incubating the egg. On the day that we found the first chicks we scooped one up, without telling Mummy, and brought it home. Mummy was... not thrilled.

On that same day – April 1st – we also found an egg that was hatching, in another nest. It was cracking at one end. We couldn't sit and watch it hatching, because that would not have been fair to the parents – they were already flying above us, scolding – and it would not have been good for the chick or the unhatched eggs, which would have got very hot without the mother bird to shield them from the sun. But I wanted to know how long it would take to hatch, and so I kept going back. Then, after a couple of hours, we decided to bring it on board. This second arrival was actually Mummy's idea - she said that it would be better than continually rowing ashore, every hour, and returning to the nest and disturbing the birds - but I don't think she was too delighted when we actually carried out her plan.

The mother birds did not seem at all upset at the disappearance of their young, or at any rate they did not show it.

Perhaps they can't count. They each carried on sitting on, and feeding, and caring for their remaining two young.

The kidnap victim, at one day of age.

We called the egg Remus and the chick Romulus, after the two children who grew up to found Rome. My sister is studying Latin, so she liked the names, and my mother said it was appropriate, because those two boys were brought up by wolves. It seems that I am the wolf!

To make the egg stay warm my sister put it inside her bra. This kept it at body heat. To our amazement we found that it was cheeping loudly! You could also hear it tapping and cracking. When Xoë wanted to go to bed we wrapped the egg up and put it on a hot water bottle to keep it warm.

Young seagull chicks are very pretty, being soft and fluffy. They are a similar colour and pattern to the eggs - pale brownish grey with dark brown blotches and spots. This makes it easy for them to hide amongst the stones and bushes, as they are very well camouflaged. They have pinkish grey legs and feet, and black beaks.

They do not get their yellow feet or beak until they are adult gulls, and it takes a whole four years for them to become adult. Of course, they can fly and look after themselves long before that - but at this stage we did not know how long it would be before our new pets became independent. Mummy was quite worried about it, as she was afraid that they would not want to leave home for a year or more. I was hoping that she might be right.

Remus never actually hatched from his egg. I think that we moved it about too much, so he got confused and couldn't tap in the right place. He made a little hole in the blunt end of the egg, and we could see his beak, but then he stopped tapping. We

Remus chipped a hole in the end of his egg.

could see that he was getting weaker, although he carried on cheeping endlessly. At the time we didn't know how long it was supposed to take for him to get out of the egg, although we knew that Romulus had taken less than a day. (We knew this because one day his egg was whole and undamaged, and the next day there was a fluffy chick in place of the egg.)

After more than a day of watching Remus fail to make the hole any bigger we decided to help him. Mummy took a pair of nail scissors and began to snip, very carefully, at the membrane which was around the hole, making the hole larger. The membrane had dried out and was stuck to the chick, probably because of the hot water bottle. We soaked it off with warm wet tissues. We could see that there were blood vessels on the inside of the egg shell. I found out later that all the eggs have this, but at the time we were rather worried and thought that the chick was only half formed.

Helping Remus out of his egg.

This fear was added to when, finally, Mummy tipped Remus out onto her hand and we saw a big yellow bump at the end of the chick. At first glance we thought it was the yolk, and that he hadn't finished forming (in which case he would die) but it turned out to be the chick's tummy button, all swollen up. We thought that it was because we hadn't kept the chick damp enough, but later I discovered that they are all like that when they are newborn. In fact, I also found a newborn baby blackbird which had a similar "tummy button" looking like a shrunken egg yolk.

When we got Remus out he was tiny, wet, cold, and very weak. We all thought that he was going to die. He had even stopped cheeping. I sat with him on my lap in the sunshine, to warm him up, and we fed him sugar water with a little syringe. And he

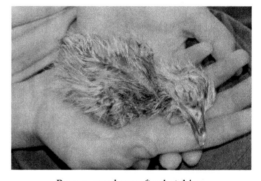

Remus, one hour after hatching.

lived! For about a day afterwards his tummy button was swollen and he had trouble with his bowel movements. But we put warm wet tissues on his bottom, and he loosened up and was cured.

Now I had two seagull chicks to study closely, and I knew that I would be able to learn far more than if I just visited the birds in their nests. I started keeping a diary to record my observations of Romulus and Remus and their relations on the island. What follows is based on my diary and so it is written chronologically, with only a few comments added afterwards.

April 16th

Everybody thinks it is very funny that we have two seagull chicks on board *Mollymawk*. Even the warden from the nature reserve on Isla Grosa thinks that it is hilarious. Nobody minds us taking them, because the yellow-legged gulls are so common that the environmentalists are actually trying to get rid of some of them. They say that they kill the Audoin's chicks. The environmentalists from the government go around punching holes in the yellow-legged gulls' eggs, so that the chicks die. If they take the eggs away the mother will lay more, it is said, whereas if they addle them she does not know any better.

I actually think that the Audouin's gulls are more aggressive than the yellow-legged gulls. We have watched two Audouin's attacking each other quite viciously, and once, when we threw out pieces of bread, we saw an Audouin's gull drive away a yellow-legged gull.

Audouin's gull – an endangered species.

Romulus and Remus are doing very well. We feed them with a syringe. We feed them mulched up cat food and sardines. One of us has to open the bird's beak and the other pushes the syringe into its mouth. Its mother obviously doesn't do this for the bird – the baby bird must open its beak for itself – but Romulus and Remus won't do this for us, so at first we had to force feed them. We fed them every twenty minutes.

Feeding a one day old chick with a syringe.

After about a week they started pecking at the food on the tip of the syringe. Remus learnt this first and Romulus was rather slow, even though he is the eldest by two days. He still has to be helped.

Ever since we got them the birds have started to cheep whenever they hear Xoë or me. They definitely recognise our voices, and they don't start getting excited when anybody else talks. They get especially excited when I laugh. Maybe I sound like a seagull cackling!

Romulus was 13 days old and Remus 11 when Remus did one of the stupidest things of his little life. Mummy and I were painting pictures of the birds, and Romulus and Remus were walking around on the table, posing for us. The chicks were determined to walk off the edge of the table and kept trying to do so. Romulus had just walked off the table and landed on his back in Mummy's palette, flapping his wings. We were all laughing, except Mummy, who was a bit annoyed. I was told to banish Romulus to the box where he and his step-brother lived.

While I was lifting Romulus into the box my pencil rolled down the table. Remus waddled after it. We were all laughing at him chasing it, and we laughed all the more when he began to attack it. Then, suddenly, without any warning, Remus picked up the pencil and swallowed it! He had never in his life managed to pick up anything before. And the pencil was almost as long as his whole body!

We grabbed Remus. We held him upside down and peered up into his throat. He was very cross about this treatment and kept wriggling and screaming. We tried making him sick, but we couldn't. Nothing we could do would get that pencil out of him. Remus just sat there

Romulus (left) and Remus (right), the day before Remus swallowed the pencil.

making funny little noises. We were very upset, because we were all quite certain that our baby was going to die. That night when we went to bed we were all very sad.

II

Baby Birds Aboard

April 17th

It is the middle of April, and all over Isla Perdiguera eggs are hatching. I have a map marking 65 of the nests, but there are lots more - there are some parts of the island where I have not been for weeks - and every time I go ashore I find some nests that I have overlooked. There are still very few nests that have hatched compared to the number of nests that only have eggs. I have a collection of egg-shells from the chicks that have hatched, with the numbers of the nests written on them. They are all slightly different colours, with different patterns of spots and blotches, but the chicks are all the same colour with only a slight variation in the pattern of their spots.

The incubation period of N1 (the first nest that I found) was 28 days. I watched this nest very closely. On March 9th it had one egg. The next day it had two, and on the third day it had three. 28 days was the time from when the first egg was laid to when the first one hatched. This is interesting, because the websites all say that the incubation period is 24 to 27 days. I was not able to time the incubation at any other nests as I do not have exact dates for when their eggs were laid. Nearly all of the nests have three eggs. N11 has only two, but I think that the mother may have died. The eggs are always warm, but they are in the sun. They have not hatched yet, and I think they are overdue.

When the chicks hatch it is very hard to find them after they are more than a day old. The newborn chicks are helpless, but once they are a day old they can shuffle into the bushes on their tummies, and when they are two days old they can usually run and hide in the bushes. The eggs in N7 were the first eggs in my survey to begin hatching, but at first we thought that something had attacked the nest and eaten the eggs. On March 31st I found, to my horror, that there was only one egg left, but there was an eggshell right beside the nest and it was surrounded by dead black beetles. We could not work out what had happened.

Next day, when we went ashore in the morning, there was a newborn chick in the nest instead of the egg, and there were two other chicks nearby in the bush. Obviously, they had been hiding there the previous day and we had overlooked them!

We still do not know why the beetles were there. Perhaps one of the chicks found them and ate them, and then sicked them up. A few weeks later we found a pile of dead wood lice beside another nest.

N7 is the nest that Romulus came from. He was the chick that was still in the egg the day before. We took him because he was the smallest. We decided that he would not stand a chance in the nest. He would starve. Remus, on the other hand, would probably be the big boy in his nest and bully the others, because he was the first to hatch in his nest (N9). That, at any rate, was what we thought at the time. We assumed that it was simply a matter of the first chick out getting a head start. However, it seems that this is not the only factor. All chicks are not the same. Some are born strong and some are not. Some are clever and some are not, and some are greedier than others.

It is interesting to see the difference between Romulus and Remus. Romulus being the eldest by two days, we thought that he would be learning to feed himself first and that Remus would copy him. But no, it was Remus - the oldest of his brood - who first grew out of syringes and forced feeding and started eating by himself. It was he who first began to try to jump out of the box they live in (and so we had to buy a new, bigger box). The fact that he ate the pencil is just typical. Romulus would never have eaten it. He had not learnt how to eat by himself. Romulus - the youngest in his brood - sits in the corner of the box looking helpless. Remus is always the adventurous, mischievous one.

Of course, after he ate the pencil we all thought that Remus would die. We were very sad that night, but in the morning Remus was as full of life as ever, squeaking away in his usual way. He definitely did not seem to know that there was a pencil inside him.

Whenever we go to the island I always make a trip around my nests to check on as many as possible. It would take all day to check over the whole island, and I usually only have a few hours, so I have a route which I always follow. Whenever we leave the route and explore other parts of the island I always find more nests. Wherever we go all the seagulls in the vicinity fly up and start yelling. There are always birds standing on the high places all over the island. They start to chatter crossly as soon as we approach, and when we get near enough to be dangerous they fly up screaming. Soon all of the birds in the vicinity are on the wing, looking down on us and shouting a warning to the rest of the colony.

We are always careful to not stay near the nests for very long, because the mother must come back and sit on her eggs or they will die. When they are forming in the eggs and when they are young the chicks must be kept warm. But not too warm either. They must not sit around in the hot sun. If the mother and father are kept away from the eggs, or if they become ill and die, then the chicks will die. Either they will die of hunger, or else they will get too cold in the wind or too hot in the sun. The egg must not go above a certain temperature, and it must not be below a certain temperature for more than 20 minutes.

A gull chick over-heating. (We moved on, his mum returned, and he survived.)

The behaviour of the parent birds is different according to the contents of the nest. If it is eggs then the parents circle around above our heads anxiously, and they make a scolding chattering noise. There are always crowds of birds in the air, making a lot of noise, but you can always tell which ones are the parents. If the egg is hatching they might dive bomb you a little, but if there are chicks in the nest then the parents scream and flap their wings and dive bomb you angrily, again and again. You hear the rush of air in the birds' wings and the feet of the gull are centimetres above your head! As the bird bombs you it makes a quite frightening call. This call is quite unlike any other seagull noise. It is the howl of an attacker. You can see the fierce anger in the bird's eyes. Sometimes I have actually been hit, and it hurts quite a lot! Usually one bird is more keen to dive bomb than the other. I suppose it is the father. Sometimes the two birds collide above the nest, and then they get cross with each other, and sometimes they get angry with other birds circling nearby. They are clearly under a lot of stress.

Meanwhile, the chicks are usually trying to stay absolutely still in the bushes near to the nest, but sometimes they panic and run. Once, when we were doing my usual lap around the island, I was checking the eggs in N62 to see if they

"Get away from my nest!"

were warm (I just touch them quickly without turning the eggs or moving them at all, and I look closely to see if they are beginning to crack). I was doing this, when something caught my eye. The thing that had attracted my attention was a huge, fluffy chick which was running fast down the stony hillside about fifty metres away! It was bigger than Romulus and Remus, and we had thought they were just about the oldest chicks on the island.

I dropped my collecting bags and map and ran down the hill after the chick. It is not a good idea to do this unless you are absolutely sure that you can catch the chick, because it is getting chased further and further from its nest. I did catch it, in the esparto grass at the foot of the slope. It was enormous. It was clear that it was at least a week older than Romulus and Remus. Its wings were long and it had feathers sprouting there, whereas our two still only had fluff all over their bodies.

As I held the chick up to be photographed it sicked up the contents of its crop. Mummy suggested that this might be a defence mechanism. We had already seen an adult do this while we were looking at her young. Perhaps the birds hope that the predator will make a grab for the food - and while it does so the chick might manage to run and hide.

The mega-chick pecking the Young Naturalist.

The big chick had sicked up a great lump of stuff. We felt a bit guilty about the fact that he had lost his lunch, but we took the food away to analyse it. When we got home and studied it closely we found that it was a lump of beef steak! It was not even fatty.

Daddy said, "His mum's been hanging out at a good restaurant!"

I cut the piece of beef into little bits and fed it to Romulus and Remus. It was the first solid food they had ever seen and they ate it eagerly. Until Remus ate the pencil we had thought that the chicks could only manage mashed up food, dripping off the end of the syringe. They could not pick up cat food from the ground. I suppose it is too soft. They need things which are firmer. After the birds ate the beef so happily we decided to feed them on solid fish. They had always declined to eat tinned sardines, and so we bought several packets of frozen whitebait. It would be much easier for us if they would eat tinned food, because we do not have a fridge on our boat. Sometimes I manage to catch fish for the birds, but it is hard to catch fish which are small enough. When I catch bigger ones we have to fillet them and there is a lot of waste.

21st April

By the 19th of April Romulus was beginning to get little feathers on his wings. At first there were only stubby quills, but then the quills got longer and there were little tufts sticking out of them. As soon as the quills began to appear Romulus began to flap his wings, as if he were trying to fly!

Although Romulus is the more developed of the two birds his little step-brother, Remus, is now getting to be bigger.

When they were born (on April 1st and May 2nd) the chicks both weighed 50 grams.

On the 6th of April, at four days of age, Remus weighed 75g. Romulus, who is two days older, weighed 125g.

On the 11th both birds weighed 200g.

On the 12th they weighed 240g.

On the 15th they weighed 325g.

On the 17th they both weighed 400g.

But on the 20th Remus weighed 600g and Romulus weighed only 510g.

Romulus and Remus at two weeks.

Romulus has never made much noise. He was born in a nest, and perhaps because of that he knows that he has to keep quiet. He cheeps when he sees me and wants to be fed. He cheeps when he is too cold or too hot, and he cheeps if somebody picks him up (which he doesn't like). That probably sounds like quite a lot of cheeping, but we don't even notice it, and that is because of Remus. Remus does not know about keeping quiet. His mummy never had the chance to tell him. However, Remus was cheeping madly and loudly before he even came out of the egg. We have only ever come across one other cheeping egg, so it may be that his noisy behaviour has nothing to do with the fact that he has never met an adult seagull.

Remus sounds exactly like a baby's squeaky toy! For the first three weeks of his life he cheeped almost non-stop. He even used to tweet while he had a mouthful of food! He even squeaked while he was asleep! I suppose he must have stopped cheeping during the middle of the night, but I was never awake to hear if this was true. Even now, Remus still tweets for most of the day. Every now and then Mummy says, "Do shut up, Remus." But he ignores her.

"Don't worry, Mummy. Soon they will grow up and be big birds, and big birds don't tweet."

"Soon they'll learn to fly," my sister added.

"Yes," said Mummy, "And then they will make a mess all over the washing."

Mummy assumes that when they can fly, the birds will fly away. My sister did not tell her what we read on the website. There it says that immature gulls stay with their parents for four years. I don't think that the birds will stay with us for that long, but Xoë does. She always looks on the bright side.

III

Remus Lives to Tell the Tale

All this while Remus was as perky as ever and showed no signs of the fact that he had a pencil inside him. Still, I did not really think that he would continue to live. How could he live with a pencil longer than his stomach inside him? By now he was getting to be quite big. He is always hungry and he eats very, very quickly. Romulus is slow. He eats a little bit from the plate, and then I feed him some more, by hand. He does not have a very big appetite. I have to stop Remus from scoffing the lot, but although we think that Remus is huge he is not quite so big as his true brothers, on the island.

Usually I cannot find the chicks when they are more than three or four days old, but on the 22nd of April I found one hiding in the bushes near to Remus' nest (N9). Although he was heavier than Remus he didn't have such advanced quills. While I was holding him the parents got very upset. They didn't dive-bomb us, but they hovered above us looking very distressed and calling anxiously. Then one of them sicked up the contents of its crop. I suppose that food must have been intended as a meal for her baby, but she was so upset that she dropped it from the sky. I hope he found it. I wonder if her behaviour was different from the other birds because she had already had a chick stolen? After we thought of that we put the chick back very quickly and went away, so as not to upset her any more.

It seems that we need to feed Romulus and Remus more, but feeding them is already expensive. It is much more expensive than feeding the dog or the cat.

Whenever we go to the island I always visit all of my nests, and I record which ones have still got eggs and which ones have hatched. On the same day that I found Remus' brother I also checked the nests just a bit further along, near the place that we call Barbecue Beach. I found that the eggs in N8 had hatched. The chicks had gone, so I started to look for them. I was searching for them in a thick bush when I heard some rustling.
Thinking, "Here's a chick," I put my hand into the bush. But suddenly I thought, "What if it's not a chick? It could be a rabbit. Or it could be a snake."

I quickly took my hand out of the bush. Then I heard a long, loud, rattling hiss come from the bush! I scrambled back to Mummy.

We poked the bush with a stick, and it hissed again. Mummy heard it too, very clearly. She can't hear the small snakes, so it must have been a big one. We moved away and kept watch on the bush - one of us on either side - but the snake must have gone away a different way.

This snake was probably the kind known to the locals as a culebra bastarda. In English it is called a Montpellier's snake, and the Latin name is *Malpolon monspessulanus.* Although it can grow up to two metres long this snake is not especially dangerous to humans as his poison fangs are at the back of his mouth. If you are careless enough to tread on him he can't do much about it, but poking your fingers into a bush - and into the snake's mouth – would not be a good idea.

The other possibility is that it as a horseshoe whip snake. This one is a bit smaller and not dangerous at all, because it does not have a poisonous bite. I have seen both of these kinds of snake on Isla Perdiguera.

One morning when Romulus and Remus were about 22 days old and 20 days old we woke up and uncovered their box, and there was Romulus with real feathers on his wings! He looked lovely. The feathers were very dark brown and they were about 2.5 centimetres long, poking out of his bluish coloured quill coverings. The night before there had just been tufts poking out of the quill coverings. Poor Romulus didn't know what to do with his new wings. He didn't know how to tuck them up and every time they fell down he would hitch them up again. It took him all day to get the hang of it, and then you couldn't see the wings any more except when he flapped them.

Then Remus started to get his wing feathers too. Both birds had feathers appearing all over them, and for the next week they spent the whole time twitching about and preening themselves, to get rid of the fluff and the bits of quill cover. From the way they were behaving it looked as if getting feathers must be a very itchy process! There were little tiny bits of quill cover blowing around and settling all over the boat.

On April 23rd I was standing in the dinghy, pumping it, when I heard my brother shouting at the birds. The birds spend the day in the cockpit, on one of the seats, and Caesar was on deck too, using his computer. While I scrambled out of the dinghy a thought flashed through my mind - "What now? Remus has already swallowed a

Romulus struggles with his new wings.

pencil. Before that he nearly died coming out of the egg. How many lives has he got?"

But Remus was not falling in the sea or swallowing something new. He was sicking something up. That was what Caesar was shouting - "Pencil! Pencil!"

Mummy leapt up from the cabin. The pencil was half out of Remus' mouth, and the silly boy was trying to swallow it again! Mummy grabbed him and held his neck tightly, so that he could not swallow. His mouth was open, because he was yelling, and she put her fingers in and was just able to get hold of the very tip of the pencil lead and pull it out.

It all happened so quickly that by the time I was out of the dinghy Mummy was already back in the cabin waving the pencil around. It was exactly as it had looked 10 days earlier. It was lucky that it was a drawing pencil, so not terribly sharp. It still had the little rubber in the other end. While we looked at it, Remus was still outside in the cockpit cheeping loudly in protest, saying "Give me back my dinner!"

If you look closely at the photo of Remus yelling you can see the amazing construction of his mouth. There are pink things, like the teeth of a saw, in the roof of his mouth and on the floor of his mouth. These must be for holding wriggling fish. You can also see that he has a long thin tongue, and at the back of his mouth, in the roof, there is a little hole. This must be the air-hole and the place where he keeps his whistle. That explains why he can whistle while he is eating.

Since this time Remus has also swallowed a large plastic clothes peg, which fell off the washing line and into the cockpit. He ate it before anyone could get it, but he brought this up again about an hour later. He also tried to eat a nail brush, but I pulled it out of his mouth.

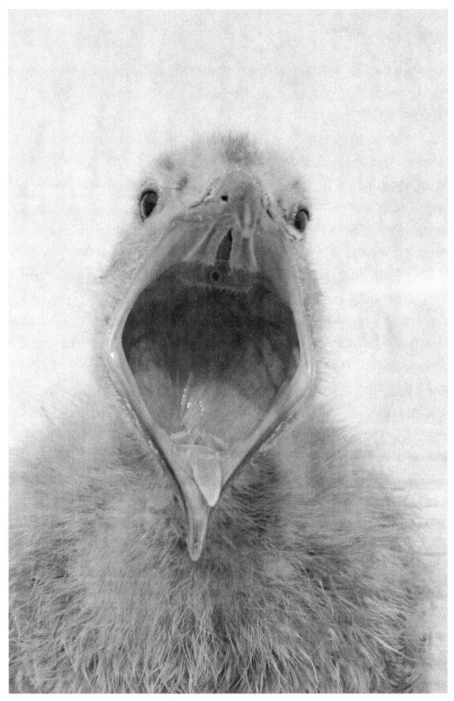

Give me back my pencil!

Would he behave like this if he lived in a nest? Do other chicks eat sticks and stones?

One afternoon Remus started panting. I thought that he was too hot. I always make sure that the birds have sunshine and some shade, but they are not always very sensible about standing in the shade. I took the birds inside, but Remus carried on panting. We decided he might be ill, and so we put the two birds in different boxes, so that Romulus wouldn't catch the illness. But it was hard to see how Remus could have caught anything. Daddy suggested that he might be thirsty - and it turned out to be true. Mummy was very puzzled, because she says the birds on the island do not have access to any water. But perhaps their parents when they regurgitate food bring them water too. We gave Romulus and Remus a big bowl of water and they both drank from it eagerly. Then they climbed into the bowl and succeeded in getting themselves completely soaked. After this we gave the seagulls a bird bath. They love it! As soon as I put the bath down they jump into it at once, even before I've poured in the water! When it is filled, they put their heads under the water and splash about. The water immediately gets filthy. It gets covered in bits of fluff and quill.

On the 27th of April Remus weighed 800g. He had become absolutely huge. Mummy joked that he was actually an ostrich chick! Within less than one month he had grown from a tiny fluffy little chick to a great big feathered bird, nearly as big as an adult. Xoë did not want him anymore, even though she helped to hatch him.

"They were so fluffy and small," she said, "and now they're just two ugly monsters." She would have liked to swap our babies for two more cute little fluff balls.

But Romulus and Remus are not ugly at all, they are beautiful! They have feathers all over them now, especially Romulus. He has a coat of pale brown and dark brown feathers, lovely big wings, and a black and white tail. But Remus still has some fluff round his neck and they both still have some fluff on their heads.

Although he is two days older and so has the best feathers, Romulus is much smaller than his step-brother. When Remus weighed 800g Romulus only weighed 650g, and when Remus kneels down, which young seagulls do a lot, he is still as tall as Romulus - or even taller!

The hand-reared chicks at one month.

On the 7th of May Romulus weighed 750g, but Remus weighed 1,000g / 1kg. It is becoming very difficult to make them stand still on the kitchen scales. We had to buy new scales that are less wobbly.

Perhaps it was that sugar-water that we gave him when he was newly hatched which has made Remus grow into such a big strong bird! Actually, Mummy and I think it is because one bird is male and one is female. It cannot just be because Remus eats more. It is true that he is very greedy, but I always make sure that Romulus gets his fair share. Mummy thinks that Remus is a male and Romulus a female. I think it is just as likely to be the other way around. In animals females are often bigger than males.

The other thing that suggests they may be male and female is the beak length. Romulus has a fairly short beak, and Remus has a longer one. After seeing this with our chicks, I started looking at birds on the island - both the chicks and the adults - and there are definitely two beak lengths, the long and the short. But which one is male and which is female?

All the eggs are hatching on the island now. We watched one of the eggs in N52 from the moment when the first crack appeared in the shell, through to the beak sticking out, and then the back sticking out, and then, finally, a tiny wet helpless chick lying in the bottom of the nest. Since the wind was blowing onshore, and making the egg cold, we had to watch this in stages. We had to make sure that the mother could return to the nest and keep her baby warm, but we went back every hour to see how things were going. It took around four hours from the time when the beak was sticking out to the time when the chick was fully hatched.

One of the other chicks in this nest was already hatched when we watched the egg. The next day we had to go away from the island for a while, and it was ten days before I could continue my survey. On that day, the 22nd of April, we found two chicks hiding in different bushes, each about 3 metres from the nest. We can sometimes watch this nest from the boat, depending on where we anchor, because it is right on the shore, and we often see the two chicks and their mother. Because we went away we do not know what happened to the third egg. Perhaps it hatched and died, or perhaps it was taken.

N52 - Day old chick and hatching egg.

We have realised that hatching from the egg is probably the most dangerous time as a chick grows up. When it is sitting with its beak sticking out, the chick is very vulnerable. Perhaps it will be too weak to get out of the egg. Perhaps it will get too hot or too cold. We have found eggs with holes in them that are alive with maggots. It is likely that the fly got in and laid her eggs on the chick after it had died, but it may not have happened that way. Perhaps if the mother left the nest a fly would get in and lay its eggs on the live chick. Maybe the maggots would eat the chick alive. Or perhaps the bite from the fly and the injecting of the eggs into the chick would kill it. We often see flies walking about on the eggs, as if they were looking for a way in.

The N52 chicks aged 1 day and 1 hour.

The biggest danger, both for hatching eggs and for very young chicks, seems to be heat. Usually on a hot day all of the mothers are keeping their chicks and their eggs in the shade, but if somebody comes close to the nest then they have to fly up. If somebody was sitting near to a nest and keeping the mother off her eggs, then they would get too hot.

Once, we were having a barbecue with some friends on the beach on the south-west side of the island. Unfortunately, we discovered half way through the meal that there was a nest nearby. We had not imagined that there would be a nest right down there, by the sand. The mother had been having a hard time. Instead of sitting on her nest she stood beside it, nervously. We checked, and one of the three eggs was hatching.

Day old chick in N99, the nest by the main beach.

The next morning we checked the nest again, and there was a chick. One of the other eggs was cracking, and the beak was already peeping out. That was a relief; they were okay. But unfortunately on that same day somebody else came and had another barbecue on the remains of ours. The parents dive-bombed them, but not very fiercely and the people would not go away. That evening we found that one of the remaining two chicks had hatched, but the third one had died in its shell, with just its beak peeping out.

We moved the picnic site, so that people will have their parties further away and the nest will not be disturbed again.

We also took the dead chick out of its shell, to have a look at it. It did not look as big-headed as Remus did when he hatched. It must have been the opposite sex to him. But it had that same big yellow ball which we thought was the yolk, when Remus hatched, but which is actually a kind of enormous belly-button. I decided to take the chick home and preserve it in alcohol, as part of my collection. Xoë was on at me about it for hours afterwards when she saw it.

"It's morbid," she told Mummy. "You shouldn't let her. I can't step out of my cabin without being confronted by a bit of dead bird. She's got a gull's wings with meat dripping off them. She's got a cormorant rotting all over the aft deck. And now you want to let her keep this corpse. I won't have it; it's like living in a graveyard."

Actually, the chick isn't doing very well. It seems to be shrivelling up.

A chick's dangers by no means stop when he is hatched. If a chick strays too far from its nest it is quite likely to be attacked by the other birds. We have found two dead fluffy chicks which had been attacked. One was stabbed in the back and the other was minus his head. We have actually seen larger chicks being attacked by adults, although they have always got away safely. Presumably they strayed too close to another nest, or perhaps there are some bullies amongst the gulls. We have also seen a sick adult being dive-bombed by one gull, just briefly. But most sick adults are not attacked.

There are also predators who might kill a chick. There are snakes… as we have seen. The snakes on Isla Perdiguera must live mostly on rabbit kittens, lizards, insects, and chicks. Once I saw a huge black snake on the beach, and recently I saw a beautiful little patterned one in the bushes. My brother has also seen a black one, in the tunnel. I have also found two snake skins, one small one and one which is huge; far longer than my height. And, of course, there was that hissing from the bush near N8. I wonder where those chicks are. I have never seen them, ever. An animal as big as the one which shed the huge snake-skin could easily manage a little chick. It could manage a large chick! Maybe it could even kill an adult gull.

IV

Swimmers and Bombers

The May Day weekend was a busy one at Isla Perdiguera, with hordes of people turning up in their boats to have a barbecue on the beach or a first swim of the season in the rather cold sea. Unfortunately, quite a few also walked around the island, and so the yellow-legged gulls nesting on the shore had a stressful time.

On May the 2nd we were walking round the island, visiting the usual nests but in the opposite order from usual. We were just coming out of the thorn bushes in the valley and still about 100 metres from the beach when Mummy started to shout.

"Roxanne! Get that chick!" she was yelling.

I looked round, and there was a chick, quite a big one, running into the sea! The sea on this side of the island was quite rough, so that he could not possibly survive. I got quite wet rescuing him, and he sicked up his lunch, which is quite usual for a chick which is in a

Rescued as he ran into the sea - but we now know that chicks can swim quite well.

panic. (It happened to be the guts of something, a young rabbit probably, or a rat. I think that gulls eat rabbits quite often.)

We continued with our inspection and when we got back round to the other side of the island Mummy went for a walk along the beach. Suddenly she was calling me again.

"Roxanne, there's another chick in the sea!"

This time she had seen the chick go into the sea when he was a very long way away. He was already a long way out. I rushed to get the dinghy - but I never found him. At the time we thought that he had sunk.

Since then we have seen many chicks swimming. I suppose it is not so surprising really. After all, ducklings can swim while they are still small and fluffy. Often, a family of two or three seagull chicks goes into the water together, and sometimes two families go together. Several times we have seen the mother and father fly out and land beside her brood. Once, three little chicks ran into the sea and swam, and their mother flew out and herded them together and led them back to the shore!

The chicks that we have seen swimming are never the very tiny ones. They seem to start at about two weeks old. At first the sea is a place to run and hide when danger is approaching. Once they are a month old then the chicks who live by the shore go swimming often, for the fun of it, and the bigger ones go a very long way out.

Mummy spends her whole time puzzling over the gulls. "Why are some of the nests out in the open, with no shelter? Why do some of the chicks run, instead of hiding?"

The question we ask ourselves most is "Why do the gulls dive bomb us?"

Obviously they are trying to drive us away from their nests.

"But it doesn't work. It has the opposite effect." says Mummy. "You're walking along, not aware of any nest being nearby, and then you suddenly start getting scolded. That's the first clue. Then when you're really near the nest they start to dive bomb you. It's a bit like that game where you have to keep asking, 'Am I getting warmer?' The nearer we get to the chicks, the more the parents dive bomb us. They actually show us where their chicks are hiding!"

Aggressive bombing betrays the fact that there is a nest nearby.

It's true. Almost half of the 107 nests that I have found have been pointed out to me by the gulls. They give you all the information. They show you where the nest is, whether there are chicks, and how old they are, from the amount of bombing. But birds are not especially evolved to cope with humans. They are evolved to deal with other animals - such as foxes. By the time a fox has come near enough to be bombed it has quite likely smelled the chicks anyway. The sooner it is discouraged, the better. I have been hit quite hard by gulls, and it hurts. I don't think a fox would like it!

Another interesting thing is the way the parents behave differently. The mother of the chicks in N9 (Remus' mother) was so upset when I picked up one of Remus' brothers that she sicked up a great gob of food. The parents of the chicks in N65 have hit me on the head, three times, while they dive-bomb me. The parents of the chicks in the beach nest (N99) do very feeble dive bombing, swooping far above my head. And the parents of N11 have never even been seen.

For a long time I thought N11 was abandoned. It was almost as old as N9 and N7 (the nests from which Romulus and Remus came) but it only had two eggs. I thought that the mother had laid two eggs and then died before laying the third. Then, on May the 2nd, we found that the eggs had hatched. But the parents still did not

Seagull chicks are irresistibly cute.

bomb us, or scold us, or pay us any attention at all. Perhaps this was because the mother had died and the father was managing to look after the chicks alone? Perhaps it is usually the mother who bombs the most. No, I don't really think so: I don't think that a single bird could look after the chicks and feed them and protect them by himself.

Some people think that gulls are aggressive, but we have watched and we think that perhaps gulls are all different characters, like people. A few gulls seem to have an instinct to attack other weaker gulls. We have seen sick birds and chicks which were swimming getting dive-bombed by other birds. Sometimes we see a chick on the island getting attacked by an adult, and often an actual fight follows, with two or three adults pulling at each other's tails, and often one bird holding another by the beak or by the neck. I suspect that this happens when a chick leaves its own territory and goes too close to another bird's nest. The gulls always drive off other gulls which come onto their territory.

We used to think that the birds dive-bombing the chicks on the water were bullies. It is never all of the birds that do it. Usually it is just one or two. Mummy now thinks it might be that they are not bullying, at all. She thinks they might be trying to encourage the juvenile, or the sick bird, to fly. I do not agree - I think these birds are being bullied - but it is true that the birds never actually seem to get hit. And sick birds sitting on the land never get mobbed at all.

Always when you walk around Isla Perdiguera you find dead seagulls. At first I thought this was perfectly natural. There is such a huge population of gulls and the old ones have to die sometime. But it is now clear that something is killing the gulls. There is some kind of illness which makes the birds weaker and weaker until they can't fly, and then they die of hunger and thirst. Their eyes become half closed and they are unable to open them properly. This could just be dehydration, or it could be part of the illness.

Six times I have found a dying bird. The first time, we took the bird back to the boat and fed it with a syringe. It got stronger - but not strong enough to fly away, and after four days it died. The second time, we left the gull where it was, and after three days it died. On a more recent occasion we took the sick bird and hid it safely under a bush, and we gave it water with a syringe. The next day it was gone, so I suppose I must have recovered and flown away.

I think that this is the best thing to do with an ill animal. If you take it away, it is sad and scared, because it has left its natural surroundings.

The photo shows a two year old bird which was smitten with the mystery illness.
Like herring gulls, yellow-legged gulls go through four phases after fledging. The one year old birds are speckled brown all over, the two year olds have grey backs but they still have brown wings, and the three year olds have just a little bit of brown on the leading edge of their wings. (This is not because they still have some of their very first feathers left. In fact, seagulls moult twice a year.) Only the fully adult birds, of four years or more, are capable of breeding.

One of the nature wardens from Isla Grosa (the nearby Audoin's seagull colony) told us that the illness which kills the gulls is botulism...! Whatever it is, the biologistas (government paid naturalists, or ecologists) have a much nastier method of getting rid of the gulls - several methods in fact. They do not want the yellow-legged gulls to nest on Isla Grosa, because they are afraid that they will attack the Audoin's, so at the egg laying time they go around punching holes in the eggs, so that the unborn chick dies. This year that still left a lot of gulls sitting around on the island, and so a few weeks ago they put little bits of poisoned bread beside a few of the nests. Apparently the parents eat the bread and then, within five minutes they fall out of the sky, dead! I suppose the chicks starve.

The biologistas are very careful with the poison. They only put it by certain nests – I suppose they choose the nests which are near to the Audouin's nests – and after the birds have eaten the bait and died they take away the corpses. This stops the poison getting into the food chain. Still, it seems rather cruel to do it at this time of year, when the chicks are around. Why couldn't they have done it earlier, when there were no chicks, instead of making them suffer?

Spot the chick. His colouration exactly matches the rocks.

We have not found very many addled eggs on Isla Perdiguera, so we do not think that the biologistas spiked any here, this year, and we do not think that they put down poison here. However, they did come and take some seagull chicks and some eggs. These are going to be raised at a zoo in the city of Murcia. Their boatman told us that the biologistas took 20 chicks and 10 eggs. This explains why some of our nests are one chick short. It is a bit of a nuisance, as it spoils my survey. I suppose that they probably took away the third chick from N52 (the nest where we watched the egg hatching).

It is getting harder to find the chicks nowadays when we go ashore. That is because most of them have hatched some time ago, and they are big enough to run and hide. Big chicks can run as fast as we can, and they can scuttle under bushes. We must walk past many chicks without seeing them. Their parents tell us that they are there, and we search, but still we often cannot see them. The big ones are in the bushes, but the little grey fluffy chicks can look just like another stone on the hillside. Twice Mummy has very nearly trodden on one!

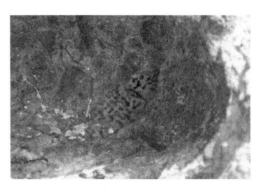
Gull chick hiding in the shadow of a rock.

Usually we only see the chicks which run, but sometimes we catch glimpses of chicks which are as big as our two - big chicks with feathers all over them. We usually only see these big juveniles while we are watching from the boat, with binoculars. Then you can usually work out who you are seeing, and which nest they come from, because the chicks stay on their territory.

The territory around each nest seems to be about fifteen to thirty metres wide. One parent is always on guard, in the middle of the territory, ready to drive off other gulls. We have noticed that it seems to be always the same parent, and we think it is probably the mother. The other bird goes away to get food, and he is usually away for hours. When he comes back the gulls both stretch out their necks and greet each other, "ka ka ka ka ka kaa kaaa." This noise is going on all the time on the island, because there are always birds returning to the nest with food, so it is a very noisy place. The parent seems to spit the food onto the ground, but it is hard to see, because it always seems to happen behind a bush!

When the chick meets the parent it bobs up and down and cheeps loudly. (Remus does this all the time whenever he sees me, but Romulus does not do it very often.) Sometimes both gulls are standing on the nest territory together, especially if the eggs are still being incubated. One bird sits on the nest, and the other one stands beside it and keeps watch very carefully. When we walk past the nests there are usually two birds bombing us, so it could be that the one who is supposed to be off getting food is often quite nearby, with his mates.

I have come to the conclusion that Remus is not so very odd after all, in swallowing a pencil. We thought that he did it because he had been brought up by humans instead of by birds, but it seems that it may be quite usual for a bird to eat whatever it finds. I have some pellets which I found beside seagull nests and which I am almost certain come from seagulls. Some contain fairly sensible things, like rabbit bones, fish bones and seeds, but one has some plastic in it, and another even has pieces of glass! I also found a pink felt pen which looked as if it had been chewed and swallowed, and once when I picked up a fairly small chick it sicked up quite a big feather.

Romulus and Remus are now great big birds with big, long wings. And to think that six weeks ago they were just tiny fluffy things! They have no fluff left on them anymore. On the 12th of May we ringed them - not with metal rings, as we don't have any yet, but with coloured cable ties. Now we have Remus-the-Red and Romulus-the-Green. In a few days time they will be flying, and we want to be sure that we can always recognise them. We do not know how far they will fly, but we have heard of six month old yellow-legged gulls travelling all the way across Spain from Murcia to Asturias, on the north coast.

Hopefully, if somebody sees a very friendly seagull with a coloured band on its leg, they will realise that it is tame and will be nice to it.

Did you spot the chick in the photo at the beginning of this chapter?

V
Flying Lessons

It is the middle of May and our six week old seagulls are no longer at all fluffy. They have feathers all over them now, and they have huge wings. To think that they were so tiny! And now that they have their beautiful wings they are desperately keen to fly.

The first flight was made on the 12th of May - but it was not really a flight, so much as a jump. Remus, the younger but bigger gull, hopped out of the box where he and his sister used to spend the night and flew down onto the floor. Next, on the 16th, the gulls both hopped up from the cockpit floor onto the seat. That was something they had never managed before, and they still could not get from the cockpit seat up onto the deck.

On the 12th of May we also took the birds swimming in the sea. Ever since they were small they have loved water. We first tried to wash their feet in it when they were rather filthy fluff balls who had walked in their dinner. We thought that we would have to hold them into the water - after all, there is no water on Isla Perdiguera, and so none of the chicks living on the hillsides there get the chance to paddle - but Romulus and Remus stayed in the water as long as we would let them, flapping, squeaking, splashing, and preening.

After that we gave them a bath every day and it got to the stage where they would recognise their bath tub amongst the pile of buckets. When I put it down near them on the cockpit seat they would run to it and leap in. Once I put it down with no water in it, and Romulus leapt in, squeaking and flapping and ducking, and trying to wash his beak. At last he stood up, cocked his head on one side (which is the only way that a seagull can look down or up) and cheeped at me, as if to say, "Where's the water? Come on, you can't fool me!"

When I took them swimming Romulus and Remus behaved rather as I would have expected them to. They were a little scared, but as it was their first time on the big, deep sea it wasn't surprising. They were used to being able to stand up in their bath. But they knew at once how to swim.

Bedraggled gull chicks with no oil on their feathers.

The gulls from the island came rushing to see what was happening. They arrived within seconds! How did they know so quickly? How did they know that some strange chicks were in the water? Did one bird who happened to be flying by raise the alarm? Within seconds the sky was speckled black with at least a hundred gulls. Romulus and Remus were afraid of the other gulls and stayed very close to *Mollymawk*. They spent a lot of time washing themselves and when we took them out again they were very soggy. They obviously did not have any oil on their feathers yet.

After we decided that Romulus and Remus were different sexes we decided to try to find out about this on the internet. We eventually found out that male gulls are larger than females. This seemed to prove that Romulus must be a girl - which was what we had already guessed. But then I wrote to someone called Cristobal, who did a study of gulls at Vigo University, and he told me that you cannot tell the difference between the sexes until they are adults. Still, we do think that Romulus is a girl, because she is so little and pretty and intelligent. Remus is big and he has a big square head, while Romulus has a rounded head, and Remus is greedy while Romulus is a delicate eater. We always think of her as a girl - and so we have decided to change her name to Romulina.

Remus is quite a pig. He wolfs down any kind of fresh fish or meat that I give to him and his step-sister. They cannot manage tinned fish because it is too soft for them to pick up. One of their favourite foods is minced meat. We have to buy it specially for them, because we do not eat any meat. They also like hake, which we buy frozen, and they like fresh fish, which is probably best for them. When I cannot catch any fish, or when we run out of frozen fish, then I give them boiled eggs, and they crumble the yolks all over the place, pecking up the tiny bits after they have swallowed the whites. They like playing with crumbs of bread and little bits of vegetable. They like a varied diet. They get bored if I give them the same thing every day.

Once we got the birds four very small mackerel. I planned to give them two each, but not at the same sitting. Remus grabbed the first fish and swallowed it whole. Then he snatched Romulina's fish - but he couldn't fit it in! He waddled about with the tail of the fish poking out of his mouth, unable to swallow it because his stomach and even his throat were already full. He did look so silly! I had to cut up Romulina's fish and hold them out to her.

On the 17th of May we watched a juvenile gull at Isla Perdiguera make what was probably his first flight. He was sitting on the water and he frantically flapped his wings, flew a very short distance just above the surface, and collapsed back into the sea. Later that same day we saw another big chick glide a few metres down the side of the island and into the sea. He was running away from us and perhaps it was his first flight. He swam a long way away, and when we looked again he was just a little brown dot sitting with a crowd of white dots - the adults. A while later we saw a juvenile flying with the adult gulls. It was flying very low over the water, and they were all yelling and some of them were mobbing it. Probably it was the same gull. Were the adults telling it to go higher? Was it a game, or was it aggression? We think the latter is most likely, because about a week later we saw a juvenile getting very badly mobbed so that it nearly fell out of the sky. However, Mummy likes to think that the adults were encouraging the juvenile and saying, "Well done!"

The following day, on the 18th, I took Romulina and Remus for another swim. This time I went in with them. The water was freezing! After the swim I picked up the birds, much to their disgust – they hate to be picked up - and I put them in the bottom of the dinghy. Romulina immediately hopped up onto the side of the dinghy, spread her wings and kicked off. Up into the air she went, about half a metre up. She hovered for a split second, wondering what to do, or how to get down, and then she flopped into the water. She had covered a distance of about a metre. This was her first ever flight. Mummy says it was like me taking my first steps! Remus stayed in the bottom of the boat, cheeping. He did not try to follow his sister.

Roxanne swimming with the gull chicks.

On this same day (17th May) we went ashore for our usual tour of the nests. We had been watching a nest which was right on the seashore, on Barbecue Beach. In fact, it was actually built on top of the seaweed which had been thrown up during a storm. The storm was on the 19th of April, so we know the nest was not there before that. It only had two eggs.

We had been watching the nest, from the boat, the day before, and the mother had been sitting there the whole time, from morning to night. She never left the nest while we were watching, and the male was standing close by for most of the time. The first thing we noticed when we got up the next morning was that the gull was not sitting on her nest. That seemed very odd, so we went ashore to look and we found that the eggs were not in the nest. One egg lay on the pebbles beside the nest, smashed open, and the other was gone. The chick inside the broken egg was dead. It looked to me as if it would have hatched in a week or so (but Mummy said she thought it looked as if it would hatch in just a couple of days.)

What happened? What did it? A human, perhaps, or an animal? Mummy's suggestion, that a wave from a passing fishing boat may have gone over the nest, was quickly ruled out. A wave strong enough to lift out the eggs would have swept away the nest - and the nest was not even wet. That only left humans or animals - but what animal living on the island could have done it?

A snake? That is not very likely. A snake would probably not come down onto the beach, and anyway a snake couldn't carry an egg away, especially not if it was being attacked by angry seagulls!

Another gull? That is even less likely, because a gull could not carry away an egg.

A rat? No, a rat would not be able to scare away the gulls. They would kill it easily.

What other predators are there on the island? We do not know of any, but there are signs. Every now and then I find a dead bird or a dead rabbit which has been torn apart. Obviously, a snake could not tear up a bird. A rat couldn't either. Rats can chew, but these birds have had their wings or their heads torn off. Another gull could not do that - and besides, I am strongly under the impression that gulls leave their dead alone, except perhaps for the long dead ones, which are only bones and feathers. Then they take some of the breast feathers to line their nests. ("Well, Darling," one says to the other, "She's dead now anyway, so we might as well make use of her feathers. I'm sure she wouldn't mind, poor old thing.")

Not many of the dead gulls are torn apart at this time of the year, and I think that this could be because the animal has found better things to eat, such as chicks - and eggs. Mummy views this theory with amusement. "What animal could live on Perdiguera without us knowing?" she asks. "Is there a panther lurking in the tunnel?"

So that just leaves people. But what sort of person would have done this? After we saw that the eggs had been taken and destroyed,

Mummy remembered that she had heard the gulls all screaming loudly very early in the morning, just after dawn. Who would be up at that time of day? In the Mar Menor, only a fisherman would be up so early, and they very rarely go ashore. Anyway, why would a fisherman want to smash the eggs? I do not understand why, but it seems that some people get pleasure from killing feeble helpless animals. Once we found a chick - a fluffy little two day old thing, so cute and lovely - which had been hung up in a tree between two branches. Only a human could do that.

Once, while we were sailing into the anchorage, we saw a family of six - mother, father and four children, the youngest aged about ten - who were climbing about on the hillside, banging the bushes with sticks and throwing stones up into the air at the gulls who were dive bombing them. They eventually found a chick and took it back to their boat. Then the boys and the man went back to look for a second chick, and they eventually found a rather bigger one. Of course, I can hardly object to other people taking away chicks, but it did not look as if these were the sort of people who would take good care of a baby seagull. Looking after a baby seagull is hard work and when Romulina and Remus were small most of my day was filled with feeding and cleaning. Would these people manage to get the baby birds to eat? Would they know how to feed them using a syringe? Would they be willing to spend lots of money on fresh fish? And would somebody who throws stones at adult birds be kind to a scruffy fledgling? Even if these chicks survive I think they will end up being cage birds. The people who took them probably live in an apartment block, like most people in Spain. How would they ever let the birds go flying? Unless you live right beside the sea you cannot really let a young gull go flying. It would be dangerous to let one loose in a town. Really, if you want to bring up a baby seagull, you need to live right beside a seagull colony.

I suppose that the sort of person who hangs baby chicks or throws stones at adult gulls is the sort of person who would also smash eggs. It is a mystery that we shall never solve - until we find an egg-eating snake or a panther that lives in the tunnel!

The day after we saw the people take away the two chicks, a friend of ours came to visit the island with his girlfriend and her two children. The children are boys aged eight and four, and they are a bit wild. When they saw that we were looking at the seagull nests they wanted to join us, but at first we tried to keep them away. We were afraid that they would step on the chicks or scare them into running away from their nests, and so we hurried away up the hill and lost the little boys. Then we thought about it some more, and we decided that we did not want Gonzalo and Jorge to grow up to be people who hang chicks or throw stones at birds. So in the afternoon we took them with us and showed them some nests with eggs and some little chicks. We even managed to show them a nest with a chick hatching. And we explained that you mustn't keep the adult birds away from the nests or the babies will die.

It worked very well. The younger boy loved the little soft chicks, and now the older boy cannot wait to visit the island again. I gave him an empty eggshell and he took it to school to show to his class.

Introducing a friend to the seagulls.

I have not yet explained where we kept our baby seagulls. It would have been nice to let them wander about wherever they liked, but that obviously was not possible. For a start, Remus would have come across all sorts of unsuitable things to eat. Also, Mummy would not have been happy if they made a mess on the beds, or even on the cabin floor. She once met a man who lived alone on an island covered in penguins, and he let the penguins wander in and out of the house – but she did not want to live in that sort of mess.

When they were tiny Romulina and Remus lived in a big plastic crate, and when they outgrew the crate we let them live in the cockpit. At night we brought them in and put them in a big plastic box. As they got bigger still their box became very cramped and we decided that they were strong enough to sleep outside on the side deck, behind a wind break. For the first two nights they were a bit frightened and kept calling me to come and get them in!

The deck was a good place for the birds while they could not fly, but when they reached the half way stage and could nearly fly it was not so good. They might have flapped their wings so much that they went up into the air, got carried off by the wind, and not been able to get back aboard.

Whenever there was someone on hand to watch them, the birds had the run of the deck.

So we decided that until they could fly perfectly we would have to put a net over the top of the side deck. When we were on deck they could come out and wander around on the whole deck, but if we were not there they had to go back on the port side deck with the net over them. Even after they learnt to fly we used to put the birds into their cage at night. We were afraid that if the wind changed and we had to move the boat in the night then our feathered friends might get left behind.

We were not at Isla Perdiguera when Romulina flew her first proper flight. We were at the other end of the Mar Menor, at La Ribera.

Perdiguera is not very far away as the seagull flies, and it is still in sight. Sub-adult gulls who do not have to worry about nest building come and hang around here. It is right beside the airport, with military jets and planes flying in and out - so it was a very appropriate place to learn to fly!

It was the 21st of May and when we lifted the net off the side-deck the birds hopped up onto the coach roof and both started to flap their wings. This was quite usual, but this time Romulina looked up at the sky. She had often seen adult gulls flying above the boat and from the time when she was just a few days old she had watched them. Now she wanted to be up in the sky too.

Remus was less keen. He had never paid much attention to other birds flying around the boat. He somehow knew that if he flapped enough he would fly, and he sort of wanted to, but at the same time he was scared. He flapped his wings and went up - but then he fell straight down again onto the deck, because he had stopped flapping! Then he began to flap quite desperately - but he made sure that he was hanging onto a rope with his beak!

Romulina flapped and flapped. Then all of a sudden she leapt bravely into the air. She caught her wing on the halyard which holds up the anchor ball, and I thought that she would fall, but she did not stop flapping. There was just a brief struggle to be free of the halyard, and then she was up and flying in a big circle round the boat!
"Hooray for Romulina!" cried Mummy.

Immediately all the local sub-adult gulls turned up, and our baby started to get nervous. After just one lap she started to get worried about how she would land again. She made as if to land on the boat but she did not know how. After two laps she landed in the water. This was a good idea, but she did not even really know how to do that. She did not fold her wings in quickly enough, and so there was a big splash.

Hooray for Romulina!

She had flown both her circuits at about two metres off the water. I went to fetch her with the dinghy.

Remus, although he was quick when learning to eat, was rather slow at learning to fly. On the next day Romulina improved her flying technique and went about ten times round the boat, in a low circle. After several attempts she managed to land aboard the boat. It was very difficult for her to find a way of landing without crashing into the rigging. The juveniles on the island do not have this trouble. There are no obstructions, and there is plenty of room for a crash landing. And anyway, they usually swim back to shore.

Remus was still too scared to fly. He ran up and down the deck flapping, but he kept grabbing hold of a mooring rope while he flapped. Mummy said that he was like a child who is scared to jump into the swimming pool. However, on the next day, after a lot of running up and down, flapping wildly, Remus finally flapped up into the air and flew one and a half times round the boat before landing on the water. Romulina was very pleased! She flew along with him and then, after he had landed, she kept trying to get him up again by gently dive bombing his head. "Come on, Remus! It's such fun!"

Remus' first flight.

He did try. He flapped his wings and walked along on the water. It was ever so funny! I had to fetch him in the dinghy, but the next day when the same thing happened he saw the dinghy coming and flew back onto the boat.

The birds were both just over seven weeks old when they flew. To be exact, they were both 51 days old.

VI
Gull Talk

It is time I told you about the shape of Isla Perdiguera. The island is made of four hills. Three are joined together, forming a long curved mound with summits which we call Grasshopper Hill, Mount Yellow Foot, and Lily Peak. They make a fat sort of banana shape.

Mount Yellow Foot is the biggest of the hills, and it is in the middle of the three, so of course we named it after the rulers of the island. But in July the island will be covered in grasshoppers - there are already some tiny ones around - and that is how the second hill got its name. The smaller hill is called after some plants which I think are lilies but which only grow here in the winter, when it is damp. It is not really a peak, but a sort of blunt knob, with lots of small caves in it. After Lily Peak the land comes down into a point and ends in a little piece of shingly beach where we once found a pen shell - so we call it Pen Shell Point.

On the north side of these hills, between Mount Yellow Foot and Grasshopper Hill, there is a valley which we call Palmito Valley, after the dwarf fan palm trees which grow there amongst the thorn bushes. At the bottom of the valley lies Barbecue Beach - also called Cormorant beach, because the cormorants go there in the winter.

Looking north-west, along the man-made causeway, from Dragonfly Hill. Lily Peak is on the left and Mount Yellow Foot is on the right. The main anchorage is on the left; the cove on the right is Caleta Medusa. The buildings are ruins dating from various eras.

On the other side of the island there is a bay with a long curving sandy beach, and most of the people who visit Perdiguera do not go further than this beach. The beach becomes a long sand strip which joins the biggest part of the island to the fourth hill, which I call Dragonfly Hill. This is technically a different island – on the charts it is called Isla Esparteña - but they were joined together when the long, thin beach was put here about seventy years ago. The two islands make another, littler bay, on the opposite side of the island, which we call Caleta Medusa (Jellyfish Beach). We often anchor here, depending on the wind, but most people anchor in the big bay on the other side of the island.

When we sailed back from La Ribera to Isla Perdiguera, on the 22nd of May, we anchored first in Caleta Medusa and then, after a few days, we moved to Barbecue Beach. This is not always a very good place to be anchored, but it is the best place for the birds.

In the morning when we got up and let the birds out from under their net there were already lots of juveniles sitting on the water. Mostly they were in pairs. Romulina and Remus flew around the boat a few times, and then they flew over to the other chicks and settled on the water quite nearby. It was like children meeting for the first time. They were shy. We are pretty certain that two of the other chicks must be the ones from N7 - Romulina's nest. She spent ages sitting with the other chicks on the water and sometimes trying to get them to fly, like she does Remus, by hovering just above their heads. She loves flying, but most juveniles are lazy. Or perhaps it is that they are afraid of being mobbed. Romulina seldom gets mobbed now, she flies so well. After a while she came back to the boat and practised landing on the barrel of chain on the foredeck and on a huge paint can on the aft deck. None of the other juveniles would have been clever enough to do this. They had not practised landings at all, anywhere. When they wanted to go home they all swam ashore.

We are growing to know the birds on the island, their language and their signs. When we walk along our usual route the gulls seem to know us, and they are more relaxed than they used to be. If we go right next to a nest or pick up a chick we usually get bombed, but otherwise we just get mildly scolded. The gulls land again as soon as we have passed. "Those two come almost every day," they say to each other. "Don't worry about them. They're a nuisance but they're harmless."

Dragonfly Hill and the beach, viewed from the saddle between Grasshopper Hill and Mount Yellow-Foot. The principle anchorage is to the west, on the right hand side of the photo. Caleta Medusa is also visible, with the Mar Menor and La Manga lying beyond it, in the distance.

However, if we go somewhere different from usual there is a terrible fuss. All the birds go up in a big noisy panic, and the chicks start running around instead of just sitting tight in the bushes. It is a bit upsetting.

Yellow legged gull announcing the presence of intruders.

We have noticed that the gulls have many different ways of talking. When a parent comes back to the nest territory, after flying around, the two birds greet each other by pointing their heads to the sky and screaming, "Kaaa, ka ka ka ka ka kaa!".

Usually the one returning speaks first, and if he has only been away for a short while the other one sometimes does not bother answering. But if it has been a long trip to the rubbish dump to get lunch, the second one answers, and then they both yell at the sky at the same time.

When a chick meets a parent who is returning to the family territory it bows up and down squeaking. When they are begging for food (which they usually are) they open and shut their mouths while they bob and squeak. Remus still does this almost every time I go on deck.

When we let Romulina and Remus out of their cage every morning, a single adult gull will alert the whole colony by calling. I suppose he is a sentry. He calls, "Kaa. Kaa. Kaa." The words are slower and longer than the greeting call. A small group of gulls will come flying over to see what the fuss is about, and if they go away again without making any noise then the whole colony will ignore the alert as a false alarm. But if they join in and make the same call, all together, circling over the boat, then within a few seconds the sky above us will be black with birds. I think that at times like this the parents take it in turns to stay on their nest sites, to guard the chicks, because for several minutes there is a constant stream of birds coming and going until everybody has had a look.

The adult gulls also make a chattering noise, which is their way of scolding or warning, and they make a howl as they dive bomb us. Sometimes they make a shriek. Romulina does this every time before she lands on the boat ("Look out! You're in the way!") but sometimes a gull makes this shriek for no particular reason, while it is flying round the boat, and all of the juveniles make the same shriek when they are being mobbed by an adult. ("Mummy! Mummy! Help!")

"Look out! I'm coming in to land!"

Sometimes two gulls fly around together taking it in turns to call. One calls on a fairly high note and the other one quickly answers on a lower note, and they carry on for quite a while going high low, high low, high low. We have not really worked out what this means, but I think that it is just a way of keeping contact while flying. Later on we saw juvenile gulls flying along in pairs with an adult – presumably one of their parents – and the youngster was squawking in turn with the adult's call note.

Very occasionally a gull miaows, and we do not have any idea what this means. We never hear any miaowing at the island, but the other day we heard some while we were in the port, and it really sounded like a cat that was stuck somewhere. Mummy thought at first that it was our old cat, which we left with some friends!

Gull language is quite complicated, with more different calls than we had realised. Once we heard two gulls having a conversation which went something like this:

1st Gull, on the water astern of us : "Mow" (to rhyme with cow; much shorter than the miaow call) followed by a short version of the greeting call.

2nd Gull, flying around us : "Ku-ku-ku" (a mild version of the scold), followed by one half of the high-low call, repeated four times.

1st Gull then repeated his previous remark, and 2nd Gull again said, "ku-ku-ku".

After this, the gull on the water flew up and mobbed Remus, who was fooling about nearby and looking rather like an oversized bath toy. Then he flew off with 2nd Gull.

Did the gull in the air, who could see Remus more clearly, tell the other one to come and bully him? It looked that way. It looked as if he passed a message which said, "There's a juvenile here which ought to be mobbed. And then we'll go flying together."

The juvenile gulls cannot talk properly yet. Romulina tries to communicate with Remus, but not by talking. When she wants him to fly she dive bombs his head gently, or hovers over him. Sometimes she also pulls his tail, presumably hoping that he will be so annoyed that he will get up off the water. It does not always work - but that may be because Remus is not very bright.

On the 29th of May we went ashore on Barbecue Beach, and instead of going up through the valley or south along the coast we walked north. There were 11 juveniles sitting together in a little gang off

Barbecue Beach, and our two were amongst them. The adult gulls sat separately in a very much bigger flock.

As we walked along the coast we spotted a total of 23 juveniles which, disturbed by our approach, flew down onto the sea. On the opposite side of the island we counted 21 unfledged birds which were sitting in a tight group some distance from the beach. Of course there were lots of other big chicks, which live too far from the seaside to go swimming. They were running around in amongst the bushes, trying to keep out of sight. We saw very few little fluffy chicks.

Later that same day we walked further along the north coast of the island, towards the tunnel which cuts through Lily Peak. Before long we came across a juvenile gull lying dead beside the path. It had a fishing line coming from its mouth, and the other end of the line was tangled around its feet and caught in the bushes. We have seen very few dead chicks. We had expected to see lots, because we had been told that many of them die of hunger because the adults cannot get enough food for them. This does not seem to be true. The adult gulls spend most of the day sitting about doing nothing. They are not desperately hunting for food. There is always plenty of food at the dump and they also seem to be able to catch quite a few fish.

In total we have seen about seven small fluffy chicks which were dead. One seemed to have drowned after he got stuck in a hole in a heavy rainstorm. There was so much rain that day that the roads around the Mar Menor were all flooded, and when we visited the island we expected to find that a lot of the nests and the chicks had been washed away – but actually there was no damage and only this one death. The adult gulls must have sat over their fluffy babies and kept them warm and dry.

One of the other small chicks appeared to have died of heat exhaustion – which is a much more usual problem here. One was minus his head, and one had been stabbed in the back, perhaps by an adult gull. The others we were not sure about. We have also found two dead juveniles - feathered chicks not quite ready to fly. In both cases there was a dead adult nearby, so presumably the chicks died of hunger when the parent died.

About six weeks of age, and nearly ready to fly.

This juvenile that we found beside the path had obviously swallowed a fishing hook. Further along we found another chick, slightly younger, which was lying dead on the rocks below the path, and then we found another quite near it. Then we came to a clearing in the bushes with three dead juveniles, feathered but not quite ready to fly, lying strewn on the ground. One had its tail torn off, one had its back bitten and its head bitten, and one had its wings both torn off.

To have six dead chicks within a hundred metres something must have done it. A dog, it seems, killed five of them. Then we remembered that three nights before, some fishermen had been here in the night, with their rods. I suppose they must have had a dog with them. If a dog was running loose on the island it might easily have killed five times as many chicks as we found. We only walked on the path, and quite probably there were other dead chicks in the bushes.

Chicks are well camouflaged, but a dog would be able to smell them, and in the night they would panic and run. They cannot run very fast, and when they try to run they usually end up falling over or getting stuck in bushes.

The fishermen might even have deliberately cast to the fledgling chick, I suppose. Usually a line has a weight as well as a hook, and a seagull cannot dive after the bait. It can only eat things which are near the surface. The same fishermen left their rubbish scattered all over the place - beer cans and crisp packets and empty boxes that came from the bait machine. I do not understand why people do this sort of thing. Mummy says that some of them are wrong in the head and ought to be locked up, but that most of them are just not very intelligent and have not been properly educated. We think there should be notices on the island, explaining that the gulls are trying to breed. Lots of the people who come to the island do not even realise this!

Almost all of the eggs on the island have hatched now, and most of the chicks can fly or nearly fly. But at the beginning of May we started finding new nests. The first new nest that we found was so near to N52 that at first we thought it was a new nest built on top of the old one. We wondered if a bird which had lost her first brood would try again. But the two N52 chicks were alive and well, so that did not make sense. Then we realised that it was a new nest, one bush along from N52. It is N100 on my map.

All of the new nests, with one exception, have only two eggs in them. I still think that they are probably the nests of birds who lost their first brood, but it could be that the chicks who were born late are only just four years old now, and are only just old enough to mate and breed. If they are first time parents this might explain why they are not especially good at it. Two of the nests were in rather silly places, too near to the beach (N99) and actually on the beach (N101), and none of the nests are very well guarded. We do not get bombed or even scolded when we visit them. It is very odd that there should be eggs still not hatched while most of the other birds are learning to fly!

N52 is one of our favourite nests. It is right on the edge of Caleta Medusa, on the side of Grasshopper Hill. It is only just above the water. It had three eggs, and I have described how we watched one of them hatch. Something happened to the third chick. When we left the nest that day there were two chicks and an egg which was cracking, and when we came back a few days later there were only two chicks. It is possible that the other chick fell in the sea, but I do not think that is very likely. The day after we left the island was the day when the naturalists came and took twenty five chicks for the new zoo in Murcia, and I think that the younger of the N52 brothers probably now lives there. I expect he gets lots of fish to eat and is very happy, but I hope there is a pond at the zoo because if he is anything like his siblings he will need one. They spend all day in the water, paddling around close to the shore or swimming so far away that they are just dots. After they got their full set of feathers they did not want to come out of the water at all, and their mother used to come down onto the shore to be with them. I suppose she even fed them there. These two chicks are two weeks younger than Romulina and Remus and when we were anchored here they tried to make friends, but one or other pair would always swim away before they came very close, like shy children. We do not see the N52 chicks in the cove anymore, so I think they must have swum round the corner and come to live in the little bay off Barbecue Beach. Probably they are part of Romulina's little gang.

It is very interesting to see the difference between Romulina and Remus. They are not only different in appearance but in character too.

Remus is like the pigeon that Gerald Durrell had. This pigeon was reared from a chick and it always thought that it was a human. It refused to fly. Remus would be the same if Romulina was not here to try to show him how he ought to behave. Romulina is always alert, but Remus never watches the other birds or keeps watch for danger. When he was small he even used to walk up to the dog doing his "feed me" act! Luckily, Poppy has been brought up with a hamster and then a kitten, so she knows she must leave small animals alone. She walked away from Remus. "Stupid squeaky toy," she seemed to be thinking.

Romulina eating lunch.

Romulina, on the other hand, has always been wary of the dog. She now drives the dog away by spreading her wings in a sort of threat. Nowadays, Remus does not beg from Poppy, but if her back is turned he runs up and pecks her hard! As a result, she gives the seagulls a wide berth!

Remus still tries to fit everything he finds into his huge mouth. He attacked Mummy's flip-flop, and the other day she turned around and found him drinking her tea. Afterwards she found him trying to clean up his mess with the deck brush...

Romulina is very clever and very alert, and she seems to have decided that Remus is just a silly baby who needs looking after - which is not far wrong. She swims beside him as he plays and flaps, and she even drives off any adult gulls who come and land too near for her liking. The other juveniles do not do this - when there is a crowd of juveniles sitting together on the water it is always Romulina who drives off any interfering sub-adults, while the other juveniles just try to swim away from them - but we have noticed that juvenile gulls on the island do help their parents to guard the territory. They lunge towards any adult that lands on their territory. I have also seen a gull scare a rabbit off his territory.

Although all adult gulls look more or less the same to us, we have noticed that the juveniles can always recognise their parents even when they do not make the greeting call. Sometimes when an adult lands beside the group of juveniles on the water one of them will paddle quickly towards it and sit close to it.

Romulina is usually at the head of the gang of chicks off Barbecue Beach. She and Remus seem to have adopted one particular friend, a pretty dark-headed juvenile about the same age as them, and they spend a lot of time leading him about. Once Remus swam right up to the bow of our boat with this friend. The friend was ducking his head in the water, like a sparrow in a bird bath, which is a thing that the gulls do when they are nervous.

"Remus!" I could almost hear him say, "What are we doing here? If my Mum catches us, I'm for it!"

An adult gull who had been following the chicks did not think it was a good idea at all to get so close, and she watched nervously until both of them were past *Mollymawk* and swimming away the other side. Then she flew over and joined them.

Romulina and Remus have learnt that if ever they are being harassed they need only to fly back to the boat to be safe again. Their mummy cannot fly after them and shoo away the other birds, the way some mother gulls do, but so long as one of us is on deck no adult will ever come to the boat! I think that the other gulls are rather impressed by the way Romulina and Remus are not afraid to land here.

Safe from harm, on their floating home.

VII

Feeding Habits

It is now the first week of June, and Romulina has become very good at flying. When she was just a novice she and Remus often used to get chased by the sub-adult gulls, but now they do not bother her.

Romulina has learnt how to land and take off from the boat in all sorts of different wind conditions. It is easier to take off when there is not much wind, because when it is windy the birds sometimes get carried backwards and tangle themselves in the rigging. But it is easier to fly when it is windy, because then they do not have to flap so hard, and they can swoop and flit all over the place. Romulina flits about almost as cleverly as an adult gull. Remus still lands in the middle of the deck, which, to be fair, is more than any other juvenile could do, but his sister has three favourite perches. One is a barrel of chain on the foredeck, one is a huge great can of paint on the starboard side of the poop deck, and the other is the outboard motor fastened to the rail on the port side of the poop deck. Romulina has made several attempts to start the outboard but so far she has had no more success than Daddy.

These perches are higher than anything else on the deck, so Romulina likes to sit on them and look around. She even sleeps on top of her perches during the day time, always keeping one eye half open for danger. The perches are all taller than the guardrail, so she can take off from them easily, with nothing in her way, and she flies from one to the other, around the boat, practising landing. She always lands upwind - even Remus knows, instinctively, that you have to take off and land upwind - and her favourite trick is to come sweeping round the stern, dive towards the water, and then swoop up onto the paint can.

Always, as she lands, Romulina makes her little warning call. I think it started because she always used to find Mummy standing in

Romlina with her asparagus.

the way with her camera. And whenever she lands she gets a little piece of cheese as a tasty treat. She loves cheese, and sometimes she flies from one perch to another just so that she can demand another piece of cheese. Often, when she was just learning to swoop up onto her perch, she was being pursued by a big mean looking adult, and the adults were always very cross and scolded loudly when they saw that this little chick was sitting where they could not get her and eating something! Once I gave Romulina a piece of asparagus, for fun, and she flew with it all around the boat and then back to her perch!

Romulina also likes to practice picking things up from the water. Remus plays this game with her, but in his case it is because he is hoping that the thing that they are playing with might be something he can fit in his mouth. The first time they played the game it was with a feather, and Romulina came back proudly holding it. Next time it was a cork, which was a bit alarming. If it had been Remus who found it he would have swallowed it. I am sure that when I grow up I will never be so worried by my children as I am by Remus. At least babies do not go off by themselves and get lost, or find dangerous things to swallow.

By now you are probably wondering what all these gulls on Isla Perdiguera find to eat. Gulls do not eat other gulls, even if they find them dead, but they seem to think of every other bird or animal as food. I suppose people are a bit like that too, unless they are vegetarians.

For gulls, humans are a source of food, although I don't suppose they realise that it is humans who put those mountains of food at the dump, near Cartagena. Gulls also eat fish, and we sometimes find small dead fish which have been dropped by the birds as we approach. We also find lots of dead rabbits, but we do not know whether a gull could catch and kill a rabbit. I think they could only kill a sick rabbit or a baby. Once we found two dead newborn rabbits lying beside N1 (the first nest that I found, and the nearest to the one where Romulina was born). We cannot work out who got the rabbit kittens out of their burrow. Surely a gull could not go down a burrow? And a snake would eat them underground where he found them.

Ten minutes later, when we walked back along the same path, one of the dead rabbits was gone. It was much too big to feed to a chick - all of the chicks were very tiny at the time - so it must have been eaten by a parent.

We find rabbit bones littered all over the island, together with chicken bones, and the bones from joints of meat. Once we found one back leg from a huge lizard. This was very surprising, as we have only ever seen small, stripy lizards on Isla Perdiguera. Would a seagull carry a lizard, or a lizard's leg, all the way across the water from the mainland? The lizard's leg was a beautiful shimmering bluish green. It was about 7.5 cm from the tip of the longest toe to the severed hip.

Another thing that we sometimes find is the wings from small birds, and once I found a tiny skull with a red beak. We cannot find a bird like this in the bird book, so perhaps it was a cage bird. I think that zebra finches have red bills.

Because the gulls are so dangerous not many other birds nest on Perdiguera. Any who do must nest in a very thick bush or else keep a very good guard. At least one pair of blackbirds is nesting on the island. They escape the gulls because their nest is in a thorn bush, but we have not seen any young, and it is hard to see how they could survive in the company of several hundred fierce gulls.

The stone curlews, with their huge yellow-rimmed eyes and their long legs, also nest here - or at least, we are fairly certain that they do. We hear their beautiful calling at night - Mummy says it is the most beautiful sound in all the world - and after a lot of patient watching we finally saw where they kept flying from and, from a distance, watched one run along cautiously and rush into the bushes. We will never find that nest, or the babies, and nor will the gulls. They are no better than us at scrambling through thorn bushes.

There are also lots of pairs of Sardinian warblers on the island, and we always hear their shrill scolding when we go ashore. We knew where one pair was nesting and we once had a very close encounter with one of the fledglings, which flapped across the path right under our feet. It was so tiny! I picked it up, before any of the gulls could, and put it back in the bush with its brother. Later we searched the bush to find the empty nest for my museum, but we can't find it, it is so well hidden.

There is also at least one pair of shelduck which we often see at Barbecue Beach. The gulls do not seem to mind them. They do not drive them off, they just ignore them and the ducks ignore the gulls. Mummy does not think

Rescuing a newly fledged Sardinian warbler.

they have a nest, because we almost always see both of them together, but I think they have a nest hidden in the bushes.

In the winter we see lots of crested larks on the island, and we still see a few now. We recently saw one with food in its beak, but we did not manage to find out where it is nesting. There are also common terns and little terns, but we think that we would certainly have spotted their nests when we row round the island. I suppose they are nesting at the salterns, which are not very far away as the tern flies. We also see a pair of turnstones and some little ringed plovers and quite a few kentish plovers. Once we saw a kentish plover pretending to have a broken wing, so we know that its nest must have been nearby, but we could not find it.

Kentish Plover.

Four egrets also visit the island or live here. We often see them wading along the shore near Barbecue Beach, and we once saw them clumsily perched amongst the thorn bushes on the hillside. The gulls do not go amongst the bushes, but we do not really think that there is a nest here.

We did find one nest which was a bit different from the gulls' nests. It was just a scrape in the ground, with hardly any lining of grass or feathers. The few feathers there were seemed to be seagull's feathers. There were three eggs and they were not quite like the yellow-legged gulls' eggs. They were a tiny bit smaller, and they were not so pointed. They were also a bit darker. We never saw what kind of bird was sitting on these eggs. One possibility is that they were the eggs of an Audouin's gull. We did not tell anybody about them, because if they really did turn out to be an Audouins' eggs the biologistas would probably come along and kill every yellow-legged gull on this part of the island.

We wanted to watch from the boat, to see what kind of bird was sitting on the eggs, but the weather was never right for us to anchor there, and no bird of any sort scolded us while we visited the nest. One day when we checked on this nest one of the eggs was hatching. Yellow-legged gulls' eggs always hatch in the same way, with the chick pecking a hole in the blunt end and then breaking out at that end, but this egg had a hole in one side. Later that same day there was no change - the hole hadn't got any bigger - and that evening we had to sail back to the port for some boring reason. The next time we visited the island, a few days later, all three of the eggs had gone. There were no chicks. Whoever hatched from those eggs had either run away or been eaten by the gulls. Or perhaps the eggs themselves were eaten by something.

Left : Yellow legged gull's eggs.
Right : The unidentified eggs were smaller and blunter. Note that one is cracking.
The photographer believes that it was not hatching but had been attacked.

All the chicks on Isla Perdiguera have slightly different coloured plumage, especially on their heads, but apart from these differences in colour all of them look more or less like Romulina. They are alert, quick, always glancing about, and they have a very graceful shape. They generally have a light coloured neck and head, and a dark brown back. They swim with their heads held upright. Their beaks are different lengths - the long one for the males and the short one for the females, I believe.

Romulina usually leads a group of gulls about, but sometimes she is led. She is one of the best flyers, although all of her crowd are very good now. Unlike the novice flyers they are seldom mobbed, and when she is mobbed Romulina does all the tricks and gliding and swinging off to one side that the adults use. If an adult swims too close for her liking she sees it off. If someone tries to bomb her she is no longer pushed down onto the water, which is what the adults seem to want. Instead she flies swiftly to the boat and swoops up onto her perch at the stern.

Remus is different. He is huge - bigger than any other juvenile in the group. In fact he is as big as an adult, and even bigger than some of the adults. He has darker feathers than any other chick. He does not have a graceful rounded head; his head is big and square. Although he has a neck like the other birds he does not use it. He sits very low in the water, like the unfledged chicks, and he does not look around. He just swims along squeaking. (He still squeaks most of the time.) Remus never leads the other chicks around, but they do not seem to think any the worse of him for being so different. It seems that he is wrongly formed, and perhaps that is why he could not get out of the egg without our help. Perhaps it was nothing to do with getting too dry. Certainly, getting too dry could not have spoiled his whole body shape.

Remus is different from other young gulls.

We always assumed that Remus' mad character was because he never saw his real mother, but it might be because he is wrongly formed. If we had picked a different egg and treated it in exactly the same way we might have ended up with a perfectly normal chick like Romulina.

If we had only taken Remus, and no other chick, we would have got completely the wrong idea about seagull development.

On the 29th of May Romulina and Remus spent all day on the water with their friends, off Barbecue Beach, rushing back for a plate of fish or a snack of cheese every now and then. They lost track of time and stayed with their friends until it was dusk. The other chicks started getting closer and closer to the shore, because the parents were rounding them up and leading them home. At last Romulina and Remus realised what was happening and took off to fly home to the boat, but by now it was quite dark, and they had never landed in the dark before. After trying and failing several times they went and landed on the water a few boats' lengths away. Daddy and I went to get them with the dinghy.

I took a large net, and Daddy rowed the boat while I tried to pick up the birds. They do not like being picked up – they never have liked it – and as we came near I could see that they were going to take off. So I reached out for the nearest one, who happened to be Romulina. It would have been better to get Remus, who was the least likely to be able to land in the dark.

As I picked up Romulina the rest of the gulls went up in a panic, including Remus. I calmed Romulina, who hates being held, and I gave her some cheese, for a special treat, and when we got back to the yacht I put her in the cage with a big plate of fish. She ate the fish and then paced up and down, worrying about Remus.

I was worrying about Remus too. He had disappeared into the darkness, and now he was lost. We all stood on deck calling him. We even searched with the night-vision scope, but there were no chicks on the water. Then, after a long while, I thought I heard Remus cheeping: "Cheep, cheep, cheep." Nobody else could hear it.

I was sure it was Remus, because no one else cheeps like that, continuously - but then the noise stopped and I thought, "Maybe the cheeping was coming from the island. Maybe it was just a baby."

But then the cheeping started again, and it began to move from one side of the boat to the other. Then it went away. Then it started again, much louder - and suddenly Remus appeared out of the dark and landed, in his usual place, on the coach roof.

Remus was a long time getting to sleep that night, and in the morning he was scared to go flying. He was like a child who has fallen off a bicycle and is afraid to get back on!

On the 2nd of June we had to move to a different anchorage, to buy some food. We moved the boat at night, so that it would be less disruptive to the birds, and so that we would not have to put them in their cage during the day, but at first, when we let them out of their cage in the morning, they did not realise we had moved. They took off and flew once round the boat, in their usual way, and then they happily set off in the direction of the island!

The island is visible from the port, but the birds cannot have recognised it. They had a perfect sense of direction. Mummy believes that they have some sort of internal compass. Although we had moved four miles, the island just happened to be on the same bearing as it had been for the previous few days.

Romulina and Remus carried on flying until they were just little dots! We could not stop them! There was nothing we could do! Fortunately a whole crowd of sub-adults followed them like a pack of wolves, and started mobbing them. When they had gone about 600 metres Remus panicked and turned back. Romulina would never abandon her brother, and so she turned back too, to look after him. They eventually dropped down onto the water about 500 metres from the boat and they sat there for almost an hour. They must have been very confused. After lots of calling and coaxing they were finally persuaded to come home. For the rest of the day they stuck very close to the boat.

That afternoon the birds had settled down, and they had a lovely time playing with all sorts of things on the water. They picked them up and flew round the boat, dropped them, swooped down on them, and played tug of war over them. I was afraid that Remus would eat the things, but he did not. Then Mummy threw in a small piece of bread, and they knew the difference. Even though we have never given them bread to eat they knew that this was food, and they had a proper tug of war over that! Their playing is good practice for when they have to fish for their supper. Sometimes I throw their whitebait into the water, one at a time, and they have to catch them before they sink.

Feeding time for Romulina.

The next day we had to sail across to the other side of the Mar Menor, and although I put Remus in his cage, to keep him safe, I decided that Romulina was sensible enough to stay with the boat - and she did. She did not fly off. She just stood on the deck.

On the next day we wanted to move back to Isla Perdiguera, where the birds really belong. They had both spent all day flying around, dodging the attacks of the sub-adults, and so we said to ourselves, "They are great big birds now, not babies, and they are probably sensible enough to stay aboard. We don't need to put them into their cage."

We hoisted the sails. This caused some confusion and consternation. Then we started to wind in the anchor, which on a steel boat is a very noisy business. Romulina and Remus paced up and down nervously, and then they both took off and landed on the water nearby.

"Never mind," we said. "They will fly along behind the boat."

We got the anchor aboard and we sailed away.

Underway. Roxanne attempts to distract Romulina from worrying about the sails.

Romulina and Remus waited until we had gone about 200 metres, and then they took off from the water and flew to us. They flew round the boat, calling, and they tried to land, but everything was different. Seagulls always land into wind, and when the boat is at anchor it always lies head to wind, but now that we were sailing the wind was across the boat. To land into the wind on her usual perch Romulina had to approach the boat from the side, and that confused her. I think the birds may also have been afraid of the sails, and Caesar said the sails were disturbing the air flow. They eventually flew off to where we had been anchored and landed there, on the water. My poor babies!

So in the end we had to anchor again, and furl the genoa, and wait for the gulls to join us. After about five minutes they both flew to the boat, and then we had to put them in their cage, which Romulina does not like any more. They were not happy birds that night, sailing along - but in the morning to their delight they found that they were back home at Isla Perdiguera.

VIII
Murder Mystery

It is the middle of June, and Remus and Romulina are about ten weeks old. They are great fun now. It was hard work getting them to this stage, but it was very worthwhile. They can both fly well, and Romulina is capable of looking after herself when the sub-adults try to mob her. Actually, the sub-adults who used to chase our two babies do not bother with them now. They know it is a waste of effort.

Remus does still give us moments of worry. We have not forgotten that he once ate a pencil and a clothes peg, or that he fell off the cockpit seat, so we do not quite trust him to keep himself safe. But Romulina is fine. She is a pretty, intelligent little bird. I have done well with her. She is a success and I am proud of her. She is friendly to me still, but wary of strangers and of the dog. Remus begs from anybody, but Romulina only squeaks when she sees me. She knows that I am her mummy. This is strange really; it ought to be the other way around, because Romulina hatched in the nest and saw her real parents, whereas the first thing that Remus saw was me.

The birds spend most of the day playing with the other juvenile gulls on the water. I think that what they would really like is for me to swim with them and their friends, and for me to sleep on the beach with them at night... Apart from anything else, I am sure that the friends would certainly *not* approve of this idea!

Romulina thinks that the next best thing would be for her friends to come aboard *Mollymawk*, but no matter how much evidence they receive to show that this is perfectly safe, the other juvenile gulls have never come aboard – yet. We did have a couple of sub-adults land on deck while Remus and Romulina were chicks. At the time we thought that they wanted to rescue these two little baby birds, but we now believe that it was just as well that the chicks were in their cage.

Sometimes the other juveniles come very near to the boat. Once, when Romulina and Remus were sitting on the water at the stern, and I was throwing fish to them, a group of six other chicks inched closer and closer - but they never quite got near enough to join in with the feast.

Romulina and Remus have learnt to come when they are called. When they were small I tried training them to come when I blew a whistle, but my brother and sister complained about the noise, so I had to give that up. Then I found that the birds knew my voice, and because I always talk to them they had learnt their names. They are rather unreliable about coming. If Remus is playing with a fishing float or a piece of plastic, then he generally ignores me. However, if Romulina is on the water nearby she will come, and if she is on her aft perch, on the outboard, and I call to her from the bow, she will come at once, with a little cry.

There is one catch. If I call, "Romulina!" they both come, and if I call, "Remus!" neither comes. Neither has the faintest idea that one of them might be called Remus. They both think that they are called Romulina.

On the 6ᵗʰ of June at sunset hundreds of gulls, males and females alike, rose up from Isla Perdiguera and flew away together, heading northwards along the Mar Menor. There were less than half as many gulls as usual left on the island. The rest just disappeared over the horizon, all going within the space of five minutes, and most of them in the first minute.

Where did the birds go? Was it a flying lesson for the juveniles? At first we were afraid that Romulina and Remus had gone, but when it got dark they came and flew round the boat, calling. They spent the night away from us, but at least we knew that they were still around. Actually, we do not think that any juveniles went away with that big flock.

Did they all go fishing? Maybe, but wouldn't a gull get more fish by itself, rather than having a huge crowd of gulls come and join in, and probably scare the fish away?

When we went ashore the next morning there were far fewer birds, but by evening time they were all back. They never came back in a great crowd, like they did when they left. They must have come back in ones and twos

On the 8ᵗʰ of June the wind came up from the north, and our anchorage off Barbecue Beach was no longer very safe, so we needed to move round to the other side of the island. It was evening time, and the birds just happened to arrive aboard at the moment when we were thinking about leaving, so the easiest thing was just to put them in their cage.

We are very aware of the fact that our "pets" are actually wild birds, and we have never liked keeping them in a cage. The purpose of the cage has always been to keep them safe, rather than to keep them from being free. When they were learning to fly they could easily have hopped over the side while we were not watching and then been unable to get back aboard. Now, they were perfectly good at flying but they were still unable to feed themselves, and so we thought it best still to use the cage at night. Otherwise, if the wind got up and we needed to move the boat, Romulina and Remus would have been left behind.

By the second week of June we had stopped using the cage at night, because Romulina had stopped coming home at night. This may have been because she did not like sleeping on a cold steel deck, but we think that it is probably because she hated being shut up in the cage. After her first night away from the boat she was so tired that she slept for most of the day, and so we think that she must have rested on the water. She cannot have been sleeping ashore, because the seagulls who live there guard their territories viciously.

After we put Romulina and Remus in their cage I sat and talked to them and fed them, but they paced up and down unhappily. By the time we arrived on the other side of the island it was dark, and since it was also blowing a gale we decided to leave the birds in their cage until morning time.

In the morning it was still very windy, and we were worried about whether Remus would be able to fly in so much wind, but he and Romulina were both very agitated so we let them out. For a long time we thought that we had lost them both, but at about midday Romulina appeared alongside the boat. She was sitting on the water with another chick. When I called her she came aboard. It was quite hard for her to land in so much wind, but after she had practised for a bit she decided to show off, and she came and landed on my head!

The people in the yacht nearby could not believe their eyes! They got their binoculars out to take a closer look!

By evening time Remus had still not returned and so Daddy and I went to look for him with the dinghy. Daddy rowed, and I called, "Romulina!". We rowed towards the juveniles which were sitting in a crowd on the water, but I could not see Remus, and he did not come. It was getting dark. We were just about to give up when I heard a familiar, "Cheep, cheep, cheep!" I looked up and saw a big, bulky, long-beaked chick flying above us.

"That's Remus!"

Once Remus had worked out who we were I didn't need to call him anymore. He gradually came closer, and eventually he landed on the water astern of us. Then he swam straight towards us, begging for food in his

Remus is unmistakable.

usual way. Usually when I pick him up, he protests and struggles, but this time he just carried on begging as I lifted him aboard. His mouth was open wide, so I put him on my lap and dropped in a piece of cheese. After he had eaten several pieces he realised that the cheese was coming from a little box, so he stood up on my lap and lent forward and scoffed all of the cheese in the box. When we got back to *Mollymawk* he seemed glad to be home, but after his meal of fish he spread his wings and hopped into the air and disappeared back into the night. Perhaps he went to look for his sister.

On the 10th of June we had to go and get food for the birds. Because they hate going in their cage we decided to leave them at the island. We left early in the morning, after the birds had been fed, and we would have to get back in the afternoon to feed them again. We sailed to Puerto Tomas Maestre, which is the place where we do the internet and get water. After we had done the shopping and stocked up on frozen fish we set off back to the island.

We were about half a mile from the island when I saw a juvenile flying above the boat. There were no other birds about. Suddenly I realised that it must be Romulina!

"Romulina!" I called, and Romulina's voice answered me. Clever little bird!

She tried to land, but we had all of the sails up, and the wind was across the boat – and the boat was moving. She could not land from her usual direction, and the rigging was in the way. She made several attempts, calling out all the time, and finally she succeeded in landing on the coach-roof, where she used to land when she was still learning to fly. She stayed with us for the rest of the journey, and after we had anchored she flew away. Then Remus appeared, fed, and then left again.

Romulina is a success.

How did Romulina know that we were coming? She must have flown around the island, because the way we approached we would not have been visible from the side where she and Remus were left. She must have been looking for us. She saw our little yellow island out in the sea, and she recognised it as her home, even though it looked very different with the sails up. Gulls are much cleverer than most people would think. And their eyesight must be very good for Romulina to have been able to see us and recognise us from so far away. I hope she had not been searching all over the Mar Menor, and worrying about where her home had gone. Not many gulls have a territory which moves!

The next day we were making our tour of the island when I saw a juvenile gull standing near the edge of the water on the beach. The beach is the place where the juveniles all get together once they have fledged. It is a no-man's land - there are no nests and no territories here - and we have seen Romulina and Remus landing here, by the water's edge. We think it is probably the only place that they have ever been ashore.

The other gulls had all flown away, but this bird was dragging a wing. I crept up and caught him. It was not only his wing that was hurt. He also had a big wound and some gouges on his head. The feathers were gone, so that he was bald here. He also had a very nasty, bloody wound above his tail, on his back, but we only glimpsed this. It was obvious that he had been attacked by another gull, or gulls. It was not the sort of damage that a dog would do. He was about the right age to be learning to fly, and he had probably flapped away from his nest site on his first flight, and wandered into somebody else's territory. Or perhaps he had simply gone flying and been attacked by the sub-adults – two and three year old gulls – which chase and attack the fledglings. We were very surprised to find that a juvenile could be so badly injured by the other gulls. We have often seen fledglings being mobbed, but we have never seen one get hurt. In the attacks that we have seen, the juvenile always escapes immediately.

Mortally wounded juvenile gull.

It was clear that the injured gull would not survive on his own, and we did not know where his parents' territory might be. I wanted to bring him back to the boat, but Mummy said no, so I put him in one of the old ruins, left over from when the army used the island during the Spanish civil war, and we brought him some water and some whitebait.

As I expected, the bird was dead in the morning. He even had maggots in his back. If he died during the night, or even the day before, the maggots would not have had time to hatch from their eggs. The eggs must have been laid by the flies while the bird was still alive, and judging by their size I think that the maggots probably hatched while he was still alive. They must have been eating the chick alive.

Since this chick died we have found a total of 23 others with injured heads and wounded, torn backs. Three were still alive when we found them, but they were dead by the next day. One was found in the late evening, just before dark, struggling out of the water. It was dripping blood and quivering all over. We put it in a bush to die in peace, unmolested.

Most of the dead birds were found on the beach. They had not been intruding on another bird's territory. Only one was found near Barbecue Beach. One was so badly battered on its back that its rib cage was open!

Why do the gulls do this to each other? It is very upsetting to find young birds which have been carefully reared by their parents, and got as far as leaning to fly, being murdered by their own kind. If we had known, when Romulina and Remus were learning to fly, that the sub-adults really wanted to kill them, then we would have been even more worried. As it is, Romulina is missing the tips of two tail feathers, and I suppose that this must have happened when one of the sub-adults attacked her. We could see that the sub-adults were being horrible, but did not realise that our "pets" were really in mortal danger.

I suppose it could be said that this bullying is a form of natural selection, with the weaker birds being weeded out of the colony – but this is not really true. A big bird will always be able to kill a smaller one, even if the small one was going to grow up to be the biggest and strongest bird of all.

Sub-adult gull attacking a newly fledged juvenile.

We cannot find anything about this killing in our bird books, and we have not found any dead juveniles in previous years. Is it a new thing? Perhaps the sub-adults would usually be busy hunting for food, but nowadays there is plenty of food and so they are bored. They seem like gangs of teenagers who go around attacking smaller children. Perhaps most seagulls are nice – or normal – and it is only a few who are murderers, killing their own kind.

On the other hand it could simply be that the sub-adults see the new immature gulls as a threat. There are now hundreds of juveniles sitting on the water in flocks. There are three or four separate flocks, one off the main beach, one in Caleta Medusa, one off

Barbecue Beach, and a fourth lesser one off the north coast of the island. These large numbers are probably alarming to the sub-adults. Maybe there will be competition for food and for a territory. There is really no room for any more nests on Isla Perdiguera. We have been told that thirty years ago there were almost no nests here, and if this is true there must have been a huge increase in the gull population.

An attacking seagull is quite fearsome.

Of course, we are only assuming that it is the sub-adult yellow-legged gulls who do the killing. We have never actually seen it happen, but we have seen the sub-adults bullying the juveniles – and who else could be doing the killing?

It seems that it is best for a young gull to fledge early on in the season, while there are few other juveniles on the water and the sub-adults are not too concerned about them. Romulina and Remus were just about the first chicks to fly, and they are now too good at flying to be murdered by the older birds. Fortunately we do not have to worry about them. It is the ones who are just making their first flights that get killed.

IX

Leaving Home

On the 12th of June Xoë and I were walking around the edge of the island when we came upon two chicks standing on a rock above the shore. One was bigger than the other. He was all lovely and golden because the sun was rising and shining on him. He was probably big enough to fly, but he did not want to. He looked very nervous. Right beside him there was another chick - a tiny, fluffy thing about two weeks old. It was snuggling up to the bigger chick, as if the bigger chick was its parent or its big brother! The next day they were standing together again, and the one after that too. They must be related in some way, or they would not be on the same territory. The parents of the little one would not let the older chick come onto their territory. But they cannot really be brothers because their ages are too different.

If the nests were very close then perhaps the chicks would get to know each other. Or if the mother of the little one died, would the older chick adopt him? Surely not! Gulls are not that generous. If a chick went into the territory belonging to another pair of gulls they would be more likely to kill it than to look after it.

However, Mummy disagrees. She believes that the older gull might actually be looking after the little one. Some animals have a strong instinct to look after other baby ones. Wolves, for instance. And Gerald Durrell's hoopoe adopted his baby jays and fed them. So, it is just possible that if an orphaned chick begged from a juvenile, and tapped its beak in the correct place, creating the appropriate stimulus, then the older bird might regurgitate its food. Or the parent of the older bird, if it found the orphan in its nest, might be stimulated to feed it. I wish that we had thought of sitting at the top of the hill with a pair of binoculars and watching to see what happened.

The N100 chicks stick close to their bush.

Below the place that these two young birds were standing, and on the same days, we also saw a fluffy two or three day old chick on the water! Chicks are almost always keen to swim, but usually they take a few weeks to discover it. They do not usually swim until they have feathers. We actually know of two chicks, in N100, who live right on the edge of the water but who still do not swim. They are very unadventurous. They were five or six weeks old before they even dared to step away from the tiny shrub that overhung their nest, and we still have not seen them swimming.

The little fluff ball living on the north side of the island must be very adventurous. He seemed quite at home paddling about in the shallows. His nest must have been right on the water's edge for him to discover the sea so soon. He has a brother, and the next day when were walking around the island they both set off for the open sea. So we left quickly. We did not want the murderous sub-adult gulls to see the chicks and kill them.

Gull Chick at about 18 days of age.

At the time we thought that these two little chicks were probably the very last ones of this year – but they were not. About a month ago, in May, we found several nests which were built much later than the others. Most of them only had two eggs in them but, so far as we are aware, none of the eggs in these particular nests hatched. Some were addled and the others just disappeared. It may be that the chicks hatched and we have not found them, but we do not really think so, because we do not get scolded when we go near to these empty nests. However, since this time we have found a few very young chicks, so some of the late-breeders did succeed. These very late nests might belong to birds which lost their first brood, or they might be the nests of gulls which could not find room earlier on, or gulls which have only just become old enough to breed. Or perhaps it is the other way around. Perhaps they are nests belonging to older birds who are no longer fit enough to fight for a nesting site and no longer very fertile.

Romulina and Remus are very independent now. They have been ashore, on the beach which joins the two halves of the island, and they always spend the night ashore. They do not spend much time with us now. It is frustrating not knowing where they go. I wish I could fly, so that I could follow them!

Our two babies are also finding some of their own food now. They are less hungry, and when they come aboard they do funny coloured squits - especially Remus, who probably eats all sorts of strange things.

Romulina is less messy than Remus. She tends to fly off to do her squits in the sea. She is also less greedy than Remus. She takes a few mouthfuls of fish and then waits for it to go down before eating any more. Remus just gulps it all down and then he looks very uncomfortable and you can watch a huge lump going down his throat! Still, this is probably the best way to succeed in the competitive world of the seagulls. Romulina will have to learn to fend for herself. If she does not eat as quickly as the other birds she will go hungry.

After eating, Romulina flies down into the water and carefully washes her beak. She always does this, and Remus never does. Then she is ready for her next adventure – chatting with her friends on the beach, or playing with me and practising flying. I sometimes get the feeling that Romulina comes to the boat for company more than for food. Often she is not hungry at all, but just comes and settles down on her perch for a siesta. Even when she is sleeping she still keeps opening one eye every second to keep watch. You cannot ever creep up on her. When I call her name she always whistles in reply.

Sometimes Romulina calls to me to come out and play. She moves from perch to perch around the boat. She swerves and glides, looking straight into my eyes while she flies past me just an arm's length away. She often flies so close to the camera that her wing brushes the lens.

Once I managed to stand stroking Romulina for several seconds while she hovered beside me in mid air. She can only do it in a strong wind. Sometimes she hovers for about half a minute over my out-stretched hand eating the

Romulina attempts to land on Roxanne's hand.

bits of cheese that I am holding! Her favourite perch is now the pulpit at the front of the boat. It took her a long time to learn to land there. She will also land on my head, if I ask her to. She does not usually fold in her wings while she is on my head. She hovers and puts hardly any weight on me, but she pulls my hair, and it hurts! Sometimes she leans over my face and tries to peck my nose! I have started to wear a hat and sun glasses, to protect myself.

On the 21ˢᵗ of June Remus and Romulina came to the boat very early in the morning. Remus was being naughty, as usual. He was destroying Mummy's shoe. I got up and spoke to him, and rescued the shoe, but I did not feed them as it was too early. They flew off, and to our surprise they did not come back when we got up.

At midday I saw Remus chasing a tern. Perhaps he was just playing, or perhaps he hoped he could make it drop its fish. We have seen him lots of times since then, but he has never come back to the

Romulina likes to land on Roxanne's head - and pull her hair.

boat. Obviously he has found out how to catch fish, or found the way to the dump. "Or perhaps," says Mummy, "he has been adopted by another seagull who has lost her own chicks."
I don't think so!

Remus only ever came home to be fed, so it is not surprising, if he can find his own food, that he does not come any more. But we were not expecting him to leave so suddenly. If it were not for the fact that we have seen him since we would think that he had met with an accident, but he is very distinctive. He is bigger than any of the other juveniles, and he looks different on the water, and he makes that persistent squeaking noise, as if he had swallowed a baby's rubber toy. One evening when Romulina was aboard he flew past and the two of them were calling out to each other. I suppose she still sees him every day and plays with him. It is strange that he should be the one to leave first. He was hatched on board the boat, and he is less afraid of people, so you would think he would stay with us for longer.

All the birds are moulting, including the juveniles. This is odd, because our bird book says that only the adults and sub-adults are supposed to moult at this time of year, but the beach is covered in feathers from both adults and juveniles. However, the juveniles are only moulting their body plumage, not their flight feathers.

Another odd fact is that the birds from the different sides of the island look different. On the west side they have dark heads, and on the other side, around Barbecue Beach, they have paler heads and white necks. Romulina used to be quite easy to spot while we are anchored in the big bay on the west side of the island, because her plumage looks like the gulls from the other side where she was born. But when we were off Barbecue Beach it was sometimes quite hard to tell them apart. Now the juveniles from the different parts of the island are mixing, and there are both dark-headed and pale headed juveniles on both sides of the island.

On the 24th of June we saw Remus with a group of other juveniles. They were flying southwards, perhaps to the dump, and he was calling. I wonder if he will ever grow out of making that cheeping noise? He is not afraid of us. His chick-hood has taught him that humans cannot fly, and unlike the other gulls he is not afraid to come close to the boat. Some friends told us that when they anchored off Barbecue Beach all of the gulls went up except one juvenile.

"That'll be Remus!" I said.

Remus no longer comes when I call, but then he was never as good at that as his step-sister. It is very satisfying to have successfully reared him from egg to independent bird, and to see that he is an accepted member of the seagull colony.

X

Island of the Insects

I t is July the 1st, and summer is finally here. Isla Perdiguera is a very different place from what it was four months ago when my survey began. We do not usually have the whole anchorage to ourselves anymore, and at the weekends it is crowded. Ashore the birds are growing up. Most of them can fly, so we do not get scolded or dive-bombed anymore when we walk around on the island.

On the 24th of June the first cicada climbed up from its underground cell, split its skin, and emerged as a fully developed winged insect. I have never actually been lucky enough to see this happen, but I have read about it and I have found the empty skins which once belonged to the cicada grubs.

For most of its life the cicada lives underground, sucking the sap from roots. The amount of time it spends underground depends on the species. Sometimes it is only four or five years, but in South America it is eighteen! Here in Spain it is supposed to be about eight or ten years. Then, one summer, the cicada gets an urge to dig its way out of the ground. It makes a tunnel and then, when the ground is hot, it breaks out into the air and climbs up a plant. Sometimes it is a bush, but several of my carapaces come from a dead, dry plant, with flowers like a daisy, hardly any higher than my knees. When it gets to the top of the plant the cicada bursts out of its skin, and instead of being a grub it is now a noisy creature with huge shiny wings! This change of lifestyle must be very strange for the cicada – a bit like dying and finding you have been turned into an angel!

Spanish Cicada. It is amazing to think that these tiny creatures might be as old as the author.

Then the cicada spends two or three weeks in the sunshine, drinking sap from the trees and making an incredibly loud, shrill, zinging noise. It is really impossible to describe this noise. If you have not heard it, you would never believe how loud it is. When you get close to the bush where the cicada is sitting you almost begin to think that he has hopped inside your own head!

The sound is presumably supposed to attract a mate – but nobody is quite sure about this, because it seems that the cicadas themselves cannot actually hear the noise. It is said that they do not have ears! Fabre, the French naturalist, actually fired two great big cannons behind some cicadas and they did not even pause in their singing. However, when Daddy is grinding steel on the beach, with his big electric grinder, the cicadas always come to join in the chorus. I suppose it could be the smell which attracts them, but I am fairly sure that it is the noise.

We knew that this was the first cicada of the year, because we had been listening out for them. We can hear them easily from the boat, and when it is still and sunny the noise is the first thing we hear when we wake up. Also, you could tell that this was a brand new cicada because he still had a greenish tinge to his shiny wings. The cicadas are late this year, because it was still a bit colder than they like it. They only come out when it is really hot weather and the ground warms up.

Although cicadas are supposed to suck sap, there is not much for them now on the island. Over the past few weeks the bushes have become brown and dead-looking, because there is no water, and the leaves have fallen off them. The wild

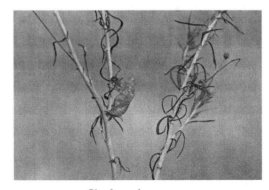

Cicada grub carapace.

fennel has gone brown, and the thorn bushes have no leaves and just a few red berries. This is good from my point of view, because it makes it easier to find nests.

I am not talking about gulls' nests. The gulls' nests have mostly fallen apart, and you can hardly see where they were, but the well woven nests of the song birds are still in perfect condition. There are a surprising number of them. I have now found five blackbirds' nests, when I thought there were only one or two pairs on the island, and I have found three Sardinian warblers' nests. I am sure that there are lots more of these. They are tiny. Smaller than a cup.

I also found a robin's nest, with two eggs inside. They were buried under the fallen leaves, so I knew they had been abandoned. They were so old that they had faded, and they were hollow. Perhaps they were not even this year's eggs. It is a long time since we have seen a robin on the island – but we once found a small brown wing, lying amongst the gulls' nests, which could have been from a robin.

On July the 10th I was searching for nests in a big patch of thorny bushes below Lily Peak. We had not explored here before. I found a big untidy nest made of thick sticks, unlike the blackbirds' and the warblers' nests, which are made of grasses, and unlike the gulls' nests, which are usually made of grasses, feathers, and leaves. It was in the middle of a big, thorny bush. It did not have a hollow in the middle of it, like most birds' nests. It was just a sort of platform. I said that it might be an egret's nest. Mummy said that in that case there probably would be more than one nest. Soon afterwards I found another similar nest a couple of bushes away, and underneath the second nest I found a small white wing feather. It is definitely not a gull's feather. So now Mummy agrees that they must be egrets' nests.

All of these birds were busy laying their eggs and raising their young right in the middle of the noisy, dangerous seagull colony!

We miss seeing the chicks when we go ashore, but we still see the juveniles sitting on the beach in a flock, and we often see them setting off away from the island in the direction of the rubbish dump.

Sometimes they go in ones and twos and sometimes as a big group, but they always have an adult to show them the way. They are not yet confident enough to go on their own as it is about ten miles each way.

It appears that Romulina knows the way to the dump too, because she is not usually hungry any more, and twice she has stayed away from the boat for almost the whole day. Quite probably she catches fish too. When I offer her food she almost always will not eat much, unless some kind of game is involved. She will usually eat her fish if I throw it in the air, or into the sea, and she will always take tiny pieces of cheese from my hand. She comes less and less every day. Soon she will leave us.

Most of the gulls have now fledged. This one will be ready in about a fortnight.

The strange murders that were happening amongst the juvenile gulls have stopped now. I think it is because there is no longer a huge flock of young gulls hanging around at the beach all day. They seem to spend most of the day away from the island, and we are not even sure if they come back at night. The birds that were killed seemed to be the ones that were just doing their first flights, and now most of the birds can fly almost as well as the adults. There are very few fledglings.

We have noticed that the few remaining fledglings do not get mobbed at all – the sub-adults seem to have moved on – so perhaps there is an advantage to being born late in the season. The worst time to be born is the busy time, from the middle of April to the middle of May. According to what we have seen, the chicks who are born then are the ones most likely to be attacked when they start flying.

Although the murders have stopped, we have still found a few more dead juveniles, and these ones seem to have been killed by the mystery illness which I wrote about earlier. The illness makes the birds weaker and weaker. First they cannot fly, and then they cannot walk. In the end they just sit on the ground with their wings propping them up, and their eyes shrivelled. They can still turn their heads, and their beaks draw a semi-circle in the sand in front of them. They die, probably because they have no food or water. The only thing that you can do for them is to feed them glucose water, with a syringe, and leave them tucked under the edge of a bush.

The mystery illness seemed to stop in the breeding season, but now it is back. Presumably it is in some way connected with the fact that the chicks have almost all learnt to fly and are getting food at the dump. This is still a bit puzzling, because although a lot of the food that we found by the nests was freshly caught fish and fish bones, we also found a lot of meat bones. If the parents feed their chicks from food found at the dump, why didn't the chicks catch the illness too?

Of course, we only find the dead chicks when they have been killed but not eaten. If any of the chicks were eaten by a snake, we would not know about it. I have now seen quite a few snakes on the island, and I am sure that they do not all live on lizards and insects. They are big snakes – and they would get much fatter eating seagull chicks.

One day in July I was walking through the tunnel with some friends. I had already told them Caesar once saw a snake in here, and as we came to the end of the tunnel we found, stretched out on the wall, a big black snake! It was at least a metre long, and it was lying perfectly still, half way up the wall, hoping we would not see it. It must have been very difficult for it to get up there. Somebody said that they could hear something cheeping in the hole a metre or so above the snake. The holes are left over from when the tunnel was made, with dynamite, and they make lovely places for the swifts to nest. You would think they would be absolutely safe up there, two and a half metres above the tunnel floor – but evidently not. The snake was obviously climbing up to try to get the chicks.

After we had stood around it, watching it, for about two minutes, the snake tried to turn round and it fell in a heap on the tunnel floor. It hissed crossly, turned around, and slithered away. It went outside the tunnel and hid under some stones.

Since this time we have seen the snake again, and I also found the skin that it had shed, lying in the mouth of a rabbit hole just outside the tunnel. It obviously manages to catch those baby birds, or else it would not spend so much

Swifts flying into the tunnel.

time in the tunnel. Probably it also tries to eat the seagull chicks from the nests outside the tunnel, but it will find this a bit harder. An adult swift would not be able to do anything to defend her young from a huge snake, but an adult seagull would certainly put up a good fight. Still, I expect some get taken.

Although most of the chicks have grown up and are flying and feeding themselves we do still occasionally come across young birds which are not quite ready to fly, and on July 5th we found a fluffy one which was still only about a week old. He was found near Barbecue Beach. We do not know if he has any brothers or sisters. We don't even know where his nest was. It must have been one of the many that were so well hidden that we overlooked them. We were very surprised to meet this little fluff ball! He was very cross when I picked him up, and the way he tried to eat his foot reminded us of Romulina and Remus when they were small and silly!

I think this little chick must be the youngest of this year's gulls. He is almost exactly three months younger than Romulina, and by the time he was born she and Remus had been flying for about five weeks. This little fellow will not be flying until the third week of August.

One of the youngest of this year's gulls.

The fluff balls that we saw swimming near the tunnel entrance now have wing coverlets, and the older chick that was looking after another youngster can fly. I am sure that there are no eggs left, and even the addled eggs have been smashed or eaten. The season for the gulls is ending, and now the insects are hatching out and taking over the island.

On July the 7th Romulina came aboard in the evening and ate a tin of mussels and an egg. We had run out of fresh food for her, and so the next morning we had to sail to the port to go shopping. We thought that Romulina would see us go. We thought she would probably be aboard when we left, but she did not come that morning. By midday we could not wait any longer and so we left. We did not manage to get back to the island until the evening of the next day. We expected that Romulina would come flying out to meet us, but she didn't. She didn't come the next day either. We have not seen her at all since the 7th.

Romulina, a week before she left us, feeding on the wing.

Almost all of the juveniles on the island have gone. There used to be a huge flock of them sitting on the beach, but now we only see a handful of late fledged birds. I suppose Romulina must have gone away with the others. We have read that young seagulls sometimes travel a very long way. Two birds ringed as chicks in the Mar Menor were found just four months and six months later on the north coast of Spain, in Asturias. Others have been found on the Atlantic coast of France, and even in the Red Sea.

Romulina and Remus have been hard work and they have been expensive, but they have been fun and they have taught us a lot. As usually happens when we study something, we have raised more questions than we have managed to answer – but we have answered one important question. Before I began my survey we were told, by the naturalists on Isla Grosa, that only one seagull chick from each clutch of three eggs would survive. That was why I thought it would be a good idea to rescue two little chicks. The naturalists had told us that the parents cannot find enough food for three hungry birds as big as themselves, and so two from each brood die of hunger. We wanted to know if this was really true – and we have found that it is not.

We have noticed that none of the adult seagulls is very busy looking for food. They spend most of the day standing around. Often they are both standing in their territory, keeping watch for intruders. When one of them flies off he does not always go to get food. Often he just goes to sit with his friends, on the beach, or on one of the ruins.

We found very few dead "fluff balls" on the island. When they are very small the biggest danger to the chicks is heat exhaustion, although being attacked by another gull is also a possibility. We only found one pair of chicks which

Seagulls on sentry duty.

seemed to have died of hunger, and we decided that their mother or father must have died first. On other occasion we also found a slightly older chick which was lying dead next to its parent.

As the birds get bigger it is very hard to tell which ones come from which nest. Often we saw five or six feathered chicks standing together on the hillside. It is a pity that we were not able to mark the chicks in some way. If we did this survey again we would put coloured rings on the chicks from certain nests - for example, red rings on all the chicks in one nest, and yellow on the three in another nest, nearby - so that we could identify them again.

But although we could not be sure that we were looking at three chicks from the same nest we knew that none of the chicks were dying of hunger. If they had been, we would have found them! As it was all the dead fledglings that we found had died of injuries.

If we were doing another survey of the seagull colony then we would definitely mark the nests, on the ground, with their numbers. We did mark five of the nests – we put a splodge of paint beside them, and painted the number in black – but we were not really sure if we ought to number them all as we did not have anybody's permission. Actually, it probably would not have mattered. Nobody would have noticed, because hardly anybody ever walks around on the island. There is no warden, because it is not a nature reserve. If anybody had noticed they would probably have thought that it was a professional survey!

Because we did not number them it was difficult to tell the nests apart and we sometimes got them mixed up. Next time we might even plot their positions with a GPS. This would work most of the time, because GPS is now accurate to just a couple of metres. Most of the nests are at least five metres apart, and many are fifteen metres apart.

Mummy has always wondered whether the gulls nest in the same place each year. "Do they fight each other for the best pitch," she asks, "or is it a case of first come, first served?"
She also thinks that the young gulls might remember the place where they were born and want to nest there – not just on the same island, or even the same

hillside, but under the same bush. Or, at any rate, on the same territory.

"Wrens nest where they were born," she says. This is probably true. There have been wrens nesting under Gran-ma and Gran-pa's back porch since Mummy was a little girl, and the ones who nest there now are probably the great-great-great-great-grandchildren of the ones who nested there when the porch was new.

"But hang on a minute," I said. "There are two birds – the male and the female – and they come from different nests. So which territory would they choose for their nesting site?"

There is another problem too – a much bigger problem. Wrens only live for about three years. Seagulls can live for about 20 years or more. If a gull rears three chicks every year from the age of four, when it becomes mature, then by the time it is 20 there will be an awful lot of gulls trying to breed under that same bush!

My brother is a mathematician, so I asked him to help me to work out how many gulls there would be after 20 years. We decided that of every three chicks, only two would live to breeding age, and we gave these survivors an average lifespan of 15 years. Then we assumed that half of the survivors were male and half female, and we decided that the gulls would nest on the territory where the female was born. This immediately halved the number, but even so, when we added up all of the chicks, and the grand-chicks, and the great-grand chicks, and so on, we found that by the time the first pair of gulls had reached the age of 15 and were ready to die, there were 69 pairs trying to nest on their territory!

Perhaps our figures might be a bit generous. It could be that only one chick from each brood reaches adulthood, and it could be that they only live, on average, for ten years. But even so, it seems clear that they do not all return to nest on the territory where they were born. They would be having to build their nests in stacks, like sky-scrapers!

If Mummy is right, and they are all trying to nest where they were born, this would explain why the colony has got so much bigger, and it might explain why the sub-adults have started killing the new fledglings. These seagulls are too successful – like people. They have run out of space.

By the time this gull is ready to nest, four years after his birth, he and his peer group will probably need to found a new colony.

We did find a few nests which were very close together. The nearest ones were three nests which were all within a metre of each other. Each one was under its own tiny bush, on the stony hillside. It might be that each pair of gulls only guarded the tiny patch beside its nest, or it could be that they all took turns to guard the bigger area around their nests. Probably there was a bit of both, with the seagulls all fiercely guarding their bush from all other gulls, including their neighbours, but only driving strangers out of the bigger territory around their bushes.

Mummy thinks that the seagulls who nested close together must be brothers and sisters. I do not really believe this. If they are related then they would probably have to be all the same age, because a chick probably never knows which of the next year's batch of juveniles is his brother. Certainly, the adults do not let the immature gulls land on their nesting territories, so they would not meet them that way.

If their population carries on getting bigger the yellow-legged gulls will have to find a new place to breed – perhaps by driving out another species, like man does – or else, like man, they will have to learn to live closer together. We have been told that the Audouin's nests are so close together that they almost touch.

Where are you now, Romulina?

Perhaps we should have taken Romulina and Remus away, to start a new colony in the Cape Verde islands! (It would not a good idea really, because they might eventually drive out another species.) It is a pity that we don't know where they have gone. I suppose the parents of the other chicks know where they are, because they probably went there themselves when they were young. I am glad that my birds were amongst the oldest, so that they could fly away with the others.

The late fledged birds have been left behind. Romulina was the best flier of them all, and Remus was the biggest, so they have had a good start in life. It is very odd not having them about anymore, and I do miss them, but to have them rejoin the seagull colony was the best thing that could have happened. It was what we always hoped for, and it all went much more smoothly than we had feared.

Yesterday Federico, the nature warden from Isla Grosa, turned up with a big coloured ring for Romulina's leg – but it is too late.

Epilogue
The Return of Romulina

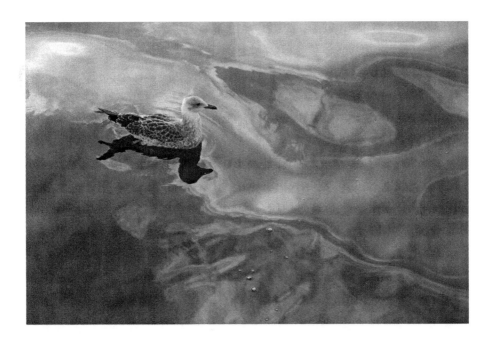

July 28th

Yesterday we were anchored at the southern end of the Mar Menor, and while we were there my sister took the dog ashore. When she came back she told us that she had seen a bird which looked just like Remus: "It didn't fly away with the others, and it was cheeping just like he always did. It had a big head…"

"No," we said. "It couldn't be Remus. He's gone away."

This evening we were sailing north, past Isla Perdiguera, when we saw five immature gulls sitting on the water nearby. Another gull came flying to join them, and I was sure that it was Remus.

"No, "said Mummy. "That gull is just like Remus, but it has a white head and a white breast. Remus had a very dark head."

At that moment one of the other gulls took off from the little group and flew towards us. She flew round the boat. I called to her and she answered. She went round and round us, trying to land on the boat. Mummy had the binoculars out.

"It can't be... The head and the breast are too white... But – yes, it is! The second tail feather from the right is missing!" (Romulina lost the tips of two tail feathers just after she fledged. I suppose she must have been attacked by a sub-adult.)

We were trying to see if the green anklet was still there, but it was almost dark now, so we could not see very clearly. Then Mummy cried, "Yes! I can see it!"

She ran to get her camera, so that we could all see the anklet in the photos. Just a while ago an immature gull came to pay us a visit, and although it did not land on the boat I thought it must be Romulina. But when Mummy showed me the photos of this bird I saw that it had a full set of tail feathers.

Romulina always had a lot of difficulty landing on the boat while we were sailing, and today she could not manage it. She and Remus swam along behind us for a while. Then he went back to join the other gulls, but Romulina tried again to land on the boat, and eventually she landed on the dinghy which we were towing. I threw her some cheese and she ate it. We decided to anchor, so that she could come aboard, but when we rounded up into the wind she flew off. She never did like the sails flapping.

Mummy looked up, and there was Romulina.

It has been three weeks since Romulina last saw us, and more than five weeks since Remus paid us any attention – but they still remember us! Perhaps they will always remember us. They are very much more intelligent than people realise.

I don't know where they have been. They must think us very boring, not being able to hop up into the air and go where we please. To them we must be as slow and lumbering as beetles. I have always thought that it must be fun to be a seagull. Now I'm sure that they have a lovely life!

August 25th

Two days after we saw Romulina and Remus again, on the July 30th, we anchored off Barbecue Beach, close to the place where our seagulls were actually born. Mummy had just done the washing and she was hanging it up. She was singing a song that is really about a girl called Serafina, but she was singing, "Romulina, oh, Romulina!" – and she looked up, and there was Romulina looking down at her! She was hovering above the deck.

When I came on deck, when Mummy called, Romulina cheeped at me and dropped down onto the water. I threw her a little bit of cheese, and she dodged aside – but then she suddenly realised what it was, and she ducked her head under the water and

Romulina is no longer a fussy eater.
She will accept almost any offering.

ate it. Then she flew up. It was quite hard for her to find a place to land, because of all the washing hanging in the way, but in the end she landed on her paint cans, on the aft deck. She was quite nervous. Her squeak sounded a bit rusty. She obviously has not used it for three weeks! But she is beginning to learn to make proper, grown-up seagull noises, and she scolded the dog just like an adult.

We have seen Romulina and Remus quite often since this day. Once they visited us when we were anchored off the port, four miles away, but I think they spend most of their time at Isla Perdiguera. There are still some other gulls living there, but most of them are adults. Perhaps our babies missed the big migration because they did not have a mother or father to show them the way. But they definitely went away somewhere, for those three weeks.

I think that the birds still go away, for a few days at a time, perhaps on a trip to some sort of feeding source. When she is visiting the island Romulina comes to see us. Once she came sailing with us for about an hour. Remus comes with her, but he will not land. He seems to have stopped cheeping, at last.

The young gulls have changed their plumage while they have been away, and now they are even more beautiful. Their heads are very white, whereas the newly fledged gulls have brown heads.

Romulina, the hand-reared but wholly independent seagull, at 4 ½ months.

Romulina is quite nervous nowadays. She lets me touch her, and she walks along the deck and peers in at the hatch, asking for food, but she is very alert. She spends the whole time turning her head to keep watch. She is also quite rough now. She still likes to play games, feeding on the wing and, when the food is all eaten, landing on my head. In fact, she even landed on my brother's head – to his alarm – and she landed on Mummy's head while Mummy was trying to take photos. But when I run out of food she pecks my fingers quite hard. No doubt her life as a seagull has taught her that she needs to fight for survival.

Romulina is also a bit scruffy now. Her feathers are ruffled, and she does not spend time tidying them. She must be having quite a hard life, but I should think that it is a good life. I quite envy her.

It is a pity that we will not be here to watch Romulina and Remus grow up. Wouldn't it be fun to come back in four years time and meet their chicks! Perhaps somebody reading this will visit the Mar Menor and will see them. I hope so.

If you do happen to see Roxanne's feathered friends, with their red and green anklets, you can drop her line using the contact form on the Schinas family's website : www.yachtmollymawk.com.

Printed in the United Kingdom by
Lightning Source UK Ltd., Milton Keynes
137550UK00001BA/256-258/P